Advances in Geographical and Environmental Sciences

Advances in Geographical and Environmental Sciences synthesizes series diagnostigation and prognostication of earth environment, incorporating challenging interactive areas within ecological envelope of geosphere, biosphere, hydrosphere, atmosphere and cryosphere. It deals with land use land cover change (LUCC), urbanization, energy flux, land-ocean fluxes, climate, food security, ecohydrology, biodiversity, natural hazards and disasters, human health and their mutual interaction and feedback mechanism in order to contribute towards sustainable future. The geosciences methods range from traditional field techniques and conventional data collection, use of remote sensing and geographical information system, computer aided technique to advance geostatistical and dynamic modeling.

The series integrate past, present and future of geospheric attributes incorporating biophysical and human dimensions in spatio-temporal perspectives. The geosciences, encompassing land-ocean-atmosphere interaction is considered as a vital component in the context of environmental issues, especially in observation and prediction of air and water pollution, global warming and urban heat islands. It is important to communicate the advances in geosciences to increase resilience of society through capacity building for mitigating the impact of natural hazards and disasters. Sustainability of human society depends strongly on the earth environment, and thus the development of geosciences is critical for a better understanding of our living environment, and its sustainable development.

Geoscience also has the responsibility to not confine itself to addressing current problems but it is also developing a framework to address future issues. In order to build a 'Future Earth Model' for understanding and predicting the functioning of the whole climatic system, collaboration of experts in the traditional earth disciplines as well as in ecology, information technology, instrumentation and complex system is essential, through initiatives from human geoscientists. Thus human geosceince is emerging as key policy science for contributing towards sustainability/survivality science together with future earth initiative.

Advances in Geographical and Environmental Sciences series publishes books that contain novel approaches in tackling issues of human geoscience in its broadest sense — books in the series should focus on true progress in a particular area or region. The series includes monographs and edited volumes without any limitations in the page numbers.

Saeid Pourmorad · Ashutosh Mohanty

Alluvial Fans in Southern Iran

Geological, Environmental and Remote Sensing Analyses

Springer

Saeid Pourmorad
Institute of Surface-Earth System Science
Tianjin University
Tianjin, China

Ashutosh Mohanty
Madhyanchal Professional University
Bhopal, Madhya Pradesh, India

ISSN 2198-3542 ISSN 2198-3550 (electronic)
Advances in Geographical and Environmental Sciences
ISBN 978-981-19-2044-8 ISBN 978-981-19-2045-5 (eBook)
https://doi.org/10.1007/978-981-19-2045-5

This Springer imprint is published by the registered company Springer Nature Singapore Pte Ltd.
The registered company address is: 152 Beach Road, #21-01/04 Gateway East, Singapore 189721, Singapore

Preface

Alluvial fans are an important and fundamental factor in many sciences such as geology, environment, natural hazards, groundwater study, and agriculture and many other related sciences and play an important and key role in these sciences. Lack of accurate knowledge of their constituent sediments has always been an important problem for experts in various sciences (Pourmorad and Jahan, 2021). In fact, alluvial fans are called cone-shaped sediments that are formed on the edges of mountains. As we go from the mountains to the plains, their thickness decreases and their width increase (Pourmorad et al. 2021).

From an economic point of view, the identification of alluvial fan deposits can be of particular importance. For example, alluvial deposits can be the center of groundwater accumulation and most groundwater reservoirs within the sedimentary basin are fed by water from alluvial deposits (Zhang et al. 2019). Three different sampling methods have been set based on different objectives and scope of the studied sites, which include sampling for granulation and petrography studies, sampling for geochemical studies and sampling for hydrochemical studies. In order to perform accurate sampling of the studied areas, the ideal points for sampling were selected after determining the accurate location of waterways and their ranking by the Strahler method (Strahler, 1952).

Most of the gold in the world is also extracted from the deposits of ancient alluvial fans in South Africa, which have remained in placer form. In addition, large amounts of uranium placer are extracted from old alluvial fan deposits in the South African sedimentary basins (Sissakian et al. 2020). What makes this book unique is that, unlike all the various books and articles published on alluvial fans, it is not limited to just one or two specific topics. In fact, this book presents for the first time a complete set of studies of sedimentology, sedimentary geochemistry, tectonics, economic geology, groundwater, geomorphology, hazards and telemetry. Another noteworthy point of this book is that, unlike many other published books, this book avoids the initial topics and theory, and all the topics discussed have been done practically to serve as a model for the study of this type of sediment around the world.

Tianjin, China Saeid Pourmorad
Bhopal, India Ashutosh Mohanty

References

Pourmorad S, Jahan SH, (2021) A model for comprehensive studies of alluvial fan deposits, Case study: Ramhormoz Mega-fan in southwest Iran. J Earth Sci Climatic Change 12(3)

Pourmorad S, Mostofa MG, Li SL, Liu CQ, Moein Z, (2021) Oil and Gas well Drilling Engineering. Lambert Academic Publishing, Germany, 480

Sissakian VK, Al-Ansari N, Abdullah LH, (2020) Neotectonic activity using geomorphological features in the Iraqi Kurdistan Region. Geotech Geol Eng 138–149

Strahler A, (1952) Dynamic Basis of Geomorphology. Geol Soc Am Bull 63, 923–938

Acknowledgments

This book arose from the work of all the paper contributors, who are working significantly in the area of an economic perspective of the identification of alluvial fan deposits importance, alluvial deposits which center of groundwater accumulation and most groundwater reservoirs within the sedimentary basin are fed by water from alluvial deposits.

We sincerely acknowledge all the professors, researchers, scientists, alluvial fan deposits and Geological Survey of Iran specialists, managers, administrators and directors without whose effort, this book could not have been written. We would like to thank the language editor at the Madhyanchal Professional University, India, for his valuable suggestions.

We would like to thank the staff at Springer for their help and support.

The authors owe a great debt of gratitude to Prof. Reza Moussavi Harami from Iran for reviewing this book with great effort and time. He generously donated his time to discuss the text's complexities and encouraged us to clarify concepts, investigate specific aspects of insight work and explain the rationales for certain recommendations.

We'd want to thank our colleagues and Geological Survey of Iron for never ceasing to inspire us to question how things are done in the corporate world. Finally, we owe a huge debt of gratitude to Dr. Ashutosh Mohanty and his University team for their invaluable editorial assistance and direction.

June 2021

Saeid Pourmorad
Ashutosh Mohanty

Book Outline

Alluvial fans are an important and fundamental factor in many sciences such as geology, environment, natural hazards, groundwater study, and agriculture and many other related sciences and play an important and key role in these sciences. Lack of accurate knowledge of their constituent sediments has always been an important problem for experts in various sciences (Pourmorad and Jahan, 2021). In fact, alluvial fans are called cone-shaped sediments that are formed on the edges of mountains. As we go from the mountains to the plains, their thickness decreases and their width increases (Pourmorad et al. 2021).

From an economic point of view, the identification of alluvial fan deposits can be of particular importance. For example, alluvial deposits can be the center of groundwater accumulation and most groundwater reservoirs within the sedimentary basin are fed by water from alluvial deposits (Zhang et al. 2020).

Most of the gold in the world is also extracted from the deposits of ancient alluvial fans in South Africa, which have remained in placer form. In addition, large amounts of uranium placer are extracted from old alluvial fan deposits in the South African sedimentary basins (Sissakian et al. 2020). What makes this book unique is that, unlike all the various books and articles published on alluvial fans, it is not limited to just one or two specific topics. In fact, this book presents for the first time a complete set of studies of sedimentology, sedimentary geochemistry, tectonics, economic geology, groundwater, geomorphology, hazards and telemetry. Another noteworthy point of this book is that, unlike many other published books, this book avoids the initial topics and theory, and all the topics discussed have been done practically to serve as a model for the study of this type of sediment around the world.

This book consists of 8 main chapters which briefly include the following:

Chapter 1: This chapter presents the generalities of the book and methods and other basic information.

Chapter 2: In this chapter, the field geological studies method and the study model of southwestern Iran are introduced. In this chapter, we have tried to introduce the most important geological formations in southwestern Iran. The introduction of these formations is important for two reasons: first, to serve as a model for other studies of alluvial fans, and second, to make the readers of this book more familiar with the names of formations and study areas, and the practical examples will be more understandable.

Chapter 3: This chapter includes advanced sedimentology studies of alluvial fans. In this chapter, sediment granulation, different sedimentary curves, sedimentary facies, structural elements, statistical parameters, estimation of ancient hydrological parameters, sediment segmentation methods, sedimentary structures and other alluvial sediment information of practical alluvial fans in southwestern Iran have been explained.

Chapter 4: This chapter includes petrographic and geochemical studies of alluvial fan sediments. This chapter first deals with the preparation and interpretation of sedimentary sections and then studies the types of geochemical studies of these sediments. The most important geochemical studies described in practice in southwestern Iran include the study of oxides, major elements, minor elements, heavy metals, weathering of sediments, ancient climate, tectonic status and mineralogical studies.

Chapter 5: This chapter includes the tectonic and morphotectonic properties of alluvial fan sediments with a practical explanation in southwestern Iran.

Chapter 6: This chapter includes environmental geological studies. In this season, environmental pollution, heavy metals, environmental indicators, groundwater chemistry and seismic hazards. In this chapter, a practical model in southwestern Iran is used.

Chapter 7: This chapter includes remote sensing studies. This chapter describes some of the methods used in alluvial sediment remote sensing studies, including the AHP method and spectroscopy.

Chapter 8: This chapter includes the interpretation and modeling of the region. In this chapter, the interpretation of the data and the final modeling of the study area are taught in practice.

This book is the result of our many years of experience as a researcher in major Iranian scientific institutes such as the Geological Survey of Iran and the Iranian Oil Company, also as a lecturer in Iranian universities and Thailand schools. The practicality of this book has made it possible for all topics to be expressed in simple language for researchers and students.

Finally, we have to be grateful to all the friends who helped me to prepare this book. We especially need to be grateful to Prof. Reza Mousavi Harami for his great help as the scientific editor of this book. We also need to thank my friends at the Geological Survey of Iran. We will also be grateful to Prof. R. B. Singh sir, who reviewed the whole paper, Ms. Sushree Sangita Dash to bring it to the Springer format and necessary corrections and Dr. Ashutosh Mohanty for contributing to the whole publication and editing process of the whole book.

For any future references and research correspondence, please contact Dr. Saeid Pourmorad.

Contents

1 Introduction . . . 1
1.1 Statement of Issue . . . 2
1.2 Importance and Necessity of Studying Alluvial Fans . . . 2
1.3 Review of Relevant Literature and Records . . . 2
1.4 Method of Studying Alluvial Fan Sediments . . . 3
1.4.1 Literature Review . . . 3
1.4.2 Field Studies . . . 3
1.4.3 Laboratory Studies . . . 5
1.4.4 Additional Studies . . . 7
1.5 Geographic Location of Alluvial Fans in Southwestern Iran . . . 8
References . . . 8

2 Geology of the Study Area . . . 11
2.1 Structural Divisions of Zagros Basin in Iran . . . 11
2.1.1 Urmia Magma Collection—Girl . . . 13
2.1.2 Sanandaj-Sirjan Region . . . 13
2.1.3 Folded Zagros . . . 14
2.2 Tertiary Stratigraphy of Zagros and the Study Area . . . 20
2.2.1 Asmari Formation . . . 22
2.2.2 Gachsaran Formation in the Study Area: . . . 22
2.2.3 Mishan Formation . . . 26
2.2.4 Aghajari Construction in the Study Area . . . 27
2.2.5 Lahbari Member in the Study Area . . . 30
2.2.6 Bakhtiari Formation in the Study Area . . . 31
2.2.7 Quaternary Sediments in the Study Area . . . 33
References . . . 38

3 Advanced Sedimentology Studies . . . 41
3.1 Granometric . . . 41
3.1.1 Histogram Curves and Normal Distribution . . . 41
3.1.2 Cumulative Frequency Curve . . . 43
3.1.3 Statistical Parameters . . . 46

3.1.4 Discussion 49
3.2 Study of Sedimentary Facies 55
3.2.1 Stone Facies or Lithofacies 55
3.2.2 Structural Elements 69
3.3 Calculation of Long-Standing Hydrological Parameters with the Help of Sedimentary Properties 74
3.3.1 Estimate the Current Strength Using the Part Size 76
3.3.2 The Thickness of the Category of Diagonal Floors and Estimating the Amount of Long-Term Drainage 77
3.4 Classification of Alluvial Fans 80
3.4.1 Ramhormoz Alluvial Fan Division (in Southwestern Iran) 80
3.4.2 Division of Alluvial Fans of Dezful 86
References 92

4 Petrographic and Geochemical Studies 95
4.1 Petrography 95
4.1.1 Petrographic Results of the Studied Areas 96
4.1.2 Petrographic Properties of the Studied Sediment Samples 99
4.2 Geochemical Studies 101
4.2.1 Oxides 103
4.2.2 Sub-Elements 103
4.2.3 Rare Earth Elements 104
4.2.4 Discussion 109
4.3 Mineralogy Studies 129
4.3.1 Illite 130
4.3.2 Chlorite 130
References 131

5 Tectonic and Morphotectonic Studies of Alluvial Fans 133
5.1 The Concept and Scope of Structural and Morphotectonic Geology 133
5.2 Tectonic Analysis and How to Study Structures in Geology 134
5.2.1 Tectonic and Morphotectonic Properties of Ramhormoz Region 135
5.2.2 Another Example of Tectonic and Morphotectonic Studies (Dezful Study Area) 142
5.3 Discussion 148
References 149

6 Environmental Geological Studies 151
6.1 Environmental Pollution 151
6.1.1 Study of Heavy Metals 152
6.1.2 Study of Environmental Indicators 152
6.1.3 Discussion 154

6.1.4 Groundwater Chemistry 156
6.2 Environmental Hazards 161
6.2.1 Seismicity Studies and Seismic Related Factors in the Studied Areas 161
References 165

7 Remote Sensing Studies 167
7.1 Primary Studies 167
7.1.1 Electromagnetic Spectrum 168
7.1.2 Factors Influencing the Formation of Satellite Imagery 169
7.1.3 Platforms 170
7.2 Analytic Hierarchy Process (AHP) 170
7.2.1 Steps of Analytic Hierarchy Process (AHP) 171
7.2.2 Flood Studies with the Help of Remote Sensing Data 172
7.2.3 Spectroscopy 183
7.2.4 Spectral Properties of Mineralogical Components 184
7.2.5 Practical Examples of Detrital Sediment Spectroscopy Studies 188
References 192

8 Basin Analysis and Modeling 195
8.1 Introduction 195
8.2 Basin Assessment 196
8.3 Sedimentary Model 198
References 199

Appendix 201

About the Authors

Dr. Saeid Pourmorad an Iranian citizen, was born in 1983 and has a doctorate degree in geology with a focus on sedimentology and sedimentary geochemistry. In addition to having a doctorate and a master's degree in geology, he has a master's degree in Remote Sensing and GIS from the best universities in Iran. His most important scientific and professional backgrounds include the following:

Thirteen years of teaching experience at Payame Noor University of Iran as a teacher; 7 years' experience for work in the Geological Survey of Iran as a senior geologist; 4 years' experience in Iran Oil Company as a wellsite geologist; 2 years' experience in the Ministry of Industry and Mines of Iran as a mine exploration expert; 1 year experience as a teacher in Thailand schools; Author of three specialized books in Persian with the names: (1) Geochemical Explorations of Sediments, (2) Geology with Help of Log Plot Software, (3) Drilling Fluids; Author of several domestic and international articles in prestigious journals; he has prepared several geological and geomorphological maps in the scales of 1:25000 and 1:10000. His main specialties are sedimentary geochemistry, petroleum geology, sedimentology and environmental studies.

He is currently pursuing a second Ph.D. in environmental sciences at Tianjin University in China.

Dr. Ashutosh Mohanty is presently working at Centre for Environment and Economic Development (CEED), New Delhi, and Professor Madhyanchal Professional University, Faculty of Science and Technology, Ratibad, Bhopal-462044, Madhya Pradesh. Dr. Ashutosh Mohanty, an International Researcher, is professor, mentor and practitioner and has 19 years of academic and professional experience in more than 15 countries with different capacities in the fields of holistic Disaster Risk Management, Strategic Emergency Policy and Risk Governance, human and institutional capacity development, Multi hazard Early warning System, Resilience and Sustainability, Humanitarian practice, Social Protection & Legal implications, IWRM, Urban Environmental Management and Climate Resilience strategies in Asia, US and Europe. He is also a Resource person and member of The Consortium for Capacity Building, CCB, INSTAAR, University of Colorado, USA, and

Co-Principal Investigator, National Geographic Explore grant USA (2020–2021). He was assigned as Director of Disaster Management and Climate Change, Shoolini University, one of the premier Himalayan universities with interdisciplinary expertise to teach Ph.D. and Master's students and prior to this, he was with Truman Graduate School-Public Affairs, Mongolia International University, and joint Graduate School Programme with University of Missouri, USA. He did his Ph.D. in Urban Environmental Risk Governance under the mentorship and research support from Dr. Michael H. Glantz, Director, Consortium for Capacity Building (CCB), INSTAAR, University of Colorado, USA, and M.Sc. in Urban Environmental Risk Management at Asian Institute of Technology, Thailand, under USAID/RUDO Fellowship. He lead a number of River Basin programmes including Amudariya Basin (Tajikistan, Afghanistan), Indus (India, Pakistan and Afghanistan), Ganges and Brahmaputra (India, Bhutan and Bangladesh), etc. He served as visiting faculty for Master's studies at the International humanitarian and social work programme, Department of Christian Social Work, Palacký University Olomouc, Czech Republic, as well as in Beijing Normal University, as faculty of Geo Sciences (2019). Ha served as Regional Capacity Development Officer at International Centre for Integrated Mountain Development (ICIMOD), Nepal, and coordinated research-based University network called Himalayan University Consortium, HUC-Center of Excellence, contributing toward sustainable mountain development among the 8 Himalayan countries (India, Bangladesh, Nepal, Pakistan, China, Bhutan, Afghanistan and Myanmar) of the Hindu Kush Himalayan (HKH) region. Being University committee team leader and international DRM export developed course curriculum for 5 universities in 4 countries along with developed Post Graduate Disaster Management courses for Odisha State Open University, Govt. of Odisha. He is associated with international research centers like University Consortium for Atmospheric Research, USA, National Center for Atmospheric Research, USA, FK-Norway, National Science Foundation, USA, and ECODIT USA. He published more than 21 International Scopus/SCI Indexed Scientific Research paper/journals and 2 Books and many are under process internationally. He coordinated, hosted and lead more than 180 Skill-based Modular Trainings and capacity development programmes with Government Officials, Community Leaders, NGOs and INGOs, and Training of Trainer (ToT) Professionals.

Chapter 1
Introduction

Abstract The alluvial fan is conical or funnel-shaped alluvium deposited at the edge of the mountains, whose thickness decreases and width increases from the mountains toward the plains (Pourmorad et al. 2021). The average slope of an alluvial fan is about 5°, but may change by more than 25° (Li et al. 2020). The radius of the alluvial fan may vary from less than a few hundred meters to more than 150 km (Tavanaei et al. 2020). Alluvial fan deposits are often red in color due to the fact that they form in an oxidizing environment (Kumar et al. 2020). The shape of these sediments depends on the tectonic and climatic conditions of the region (Bowman 2019). However, most alluvial fan sediments are formed in areas with arid and semi-arid climates with low vegetation, very low rainfall and rapid erosion. These sediments are not specific to the climate and are formed in different regions (Pourmorad and Jahan 2021). Economically, the identification of alluvial fan deposits can be of particular importance. For example, alluvial sediments can be the center of groundwater accumulation, and most groundwater reservoirs within the sedimentary basin are fed by water from alluvial sediments (Zhang et al. 2020). Most of the gold in the world is also extracted from the deposits of ancient alluvial fans in South Africa, which have remained in placer form. In addition, large amounts of uranium placer are extracted from old alluvial sediments in South African sedimentary basins (Sissakian et al. 2020). This book has a comprehensive study of alluvial fan sediments, which seeks to introduce all the sedimentary and geochemical properties of the alluvial fan. Here, an attempt has been made to study and evaluate the alluvial fan sediments of southwestern Iran as a master case with full practical application. A detailed study including sedimentology, lithology, geochemistry and morphotectonics of alluvial fans of southwestern Iran has been described in this book. These studies will contribute significantly to the identification of similar sedimentary environments and will provide a clear view of the environmental, mineral, agricultural and urban hazards of different regions.

Keywords Field studies · Sampling · Laboratory studies · Geochemical studies

S. Pourmorad and A. Mohanty, *Alluvial Fans in Southern Iran*, Advances in Geographical and Environmental Sciences, https://doi.org/10.1007/978-981-19-2045-5_1

1.1 Statement of Issue

The identification of alluvial sediments is economically important. For example, alluvial sediments serve as groundwater reservoirs and supply water within sedimentary basins. In addition, most of the world's gold is mined in the form of placer from ancient alluvial deposits such as in South Africa. In addition, large amounts of placer uranium are extracted from the sediments of ancient alluvial fans. The study of alluvial fans is also of great importance in terms of geological hazards because alluvial fans are considered as important areas of damage caused by floods, so their accurate identification is of great importance for flood prevention, environmental and agricultural sectors. Alluvial fans can also provide valuable information on neotectonics and seismic assessment of the area.

Therefore, this book aims to first study all the sedimentary properties, lithology and geochemistry of these sediments along with their origin and sedimentary environment. Then for its application, groundwater and technical studies are carried out, and finally by examining the geomorphological properties of these alluvial fans and combining the data, a general conclusion is made about the hazards and investment of the mineral sector of these sediments.

1.2 Importance and Necessity of Studying Alluvial Fans

Alluvial fans are important sources of groundwater, mineral resources such as sand and placer deposits, and help in geological hazard assessment, which is especially important for the residential areas around them. However, most of the studies that have been done on this type of sediments are either based on geomorphological properties or basic sedimentological studies, so this book shows the identification and comprehensive study of these sediments in terms of sedimentology, petrology, geochemistry and morphotectonics in each region. The comparison and examination of these regions with each other not only give a complete view of the sedimentological properties of the region but also on a practical scale help for mining purposes, agriculture, biohazard risk and drinking water supply for rural areas around them.

1.3 Review of Relevant Literature and Records

In general, studies on alluvial fans have been carried out from different perspectives, including geomorphology, geology, hydrology and tectonics, due to its importance all around the world.

The most important studies on alluvial fans include the following: the role of climate change in the development of alluvial technology in Mexicoby Dorn (2009); on the hazards of alluvial fans (Mazzorana et al. 2020); on the sedimentary properties of alluvial deposits in Yunnan Province, China (Sanchez et al. 2015; Li et al. 2020); on the assessment of groundwater resources in northwestern China alluvial fans (Xu et al. 2019); on the tectonic properties of alluvial fans California (Darcy et al. 2015; Miall 2014; Rachocki 1981) on the study of alluvial fans.

1.4 Method of Studying Alluvial Fan Sediments

The studies are based on fieldwork, land surface data and the results from geochemical and hydrochemical studies. In general, the stages of this research can be divided into four sections: data collection, field studies, laboratory studies and supplementary studies, each of which is examined separately. The following section provides an example of proposed studies for this type of sediment.

1.4.1 Literature Review

This section includes the collection of data such as relevant reports, valid articles, aerial and satellite imagery (including Google Earth imagery) and other necessary initial studies such as the collection of information from local residents about hazards and runoff in different seasons in the studied areas. The collected data were used after planning and initial analysis of planning for harvesting and field studies.

1.4.2 Field Studies

This stage of the study includes two sections: fieldwork and sample collection, each of which is examined in the following sub-sections.

1.4.2.1 Fieldworks

All the sedimentary properties of the studied alluvial fans were inspected in terms of sorting, rounding, grain size, color, texture and other sedimentary properties. In addition, Miall's (2014) naming method is used to investigate the properties, changes and lateral and vertical connections of the studied facies.

In addition to sedimentological studies, this book show studies related to tectonic and morphotectonic properties, and all tectonic features including faults, folds, anticlines, synclines, examined tectonic factors, joints, fractures and slopes.

1.4.2.2 Sampling

Three different sampling methods have been set based on different objectives and scope of the studied sites, which include sampling for granulation and petrography studies, sampling for geochemical studies and sampling for hydrochemical studies. In order to perform accurate sampling of the studied areas, the ideal points for sampling were selected after determining the accurate location of waterways and their ranking by the Strahler method (Strahler 1952). The following sections describe each.

Method

For the study conducted in southwestern Iran, ~360 sediment samples were collected from the northern regions (Dezful city) and central Khuzestan province (Ramhormoz city), southwestern Iran. About 260 samples were harvest samples, collected from the floor of three selected canals: ~1 m in Dezful region (northwest of Andimeshk), 1.8 m (northeast of Dezful) and 3.2 m (north of Dezful). The other 100 samples were collected from the floor of two selected canals in the Ramhormoz area with approximate depths of 3.8 m and 2.6 m (Figs. 1.1 and 1.2).

In these samples, the sections were selected in such a way that to a large extent, the desired sediments are included from the top to the downstream parts of these alluvial fans. The section in the Dezful region has a North–South trend, and in the Ramhormoz region the trend is from northwest to southeast. These samples were taken from the

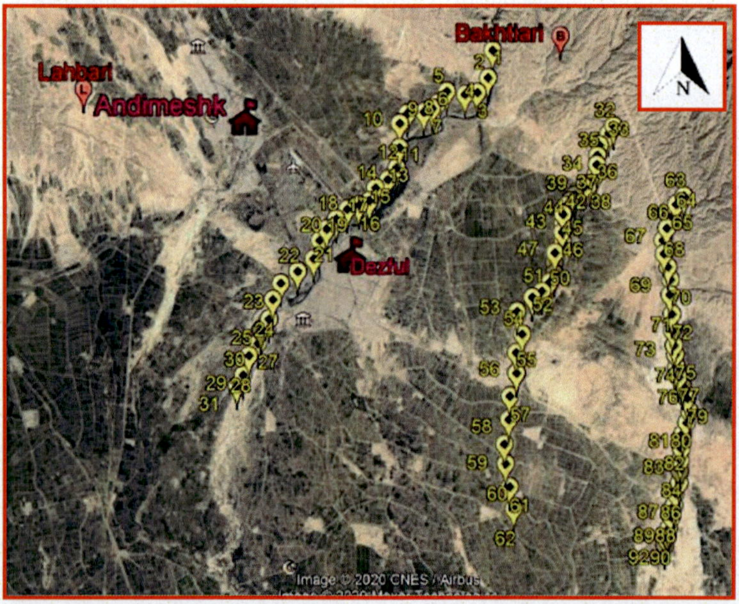

Fig. 1.1 Location of sampling points taken from the canal floor in southwestern Iran for geochemical and granulometric studies (in the north of Khuzestan province)

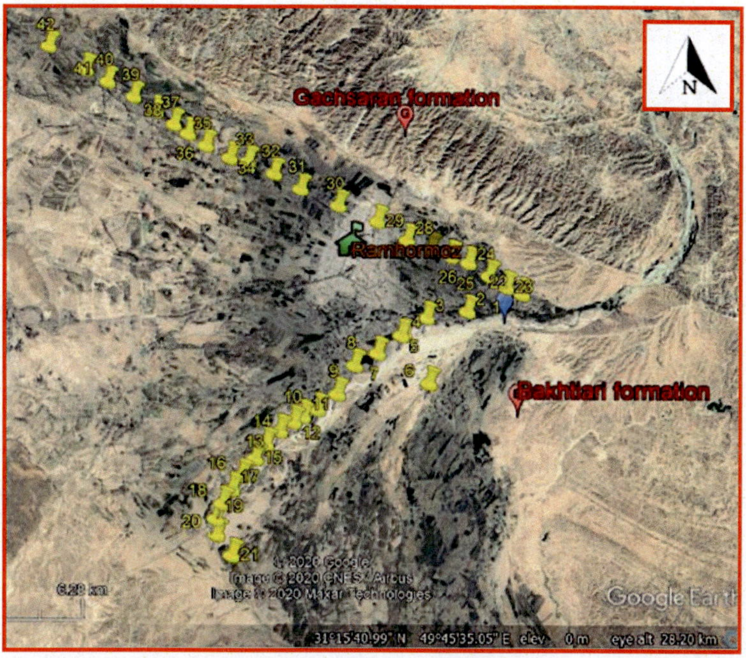

Fig. 1.2 Location of sampling points taken from the canal floor in southwestern Iran for geochemical and granulometric studies (in the center of Khuzestan province)

bottom of the canal with the help of a sample shovel with a cylindrical volume of 20 cm diameter and a depth of 25 cm. The geographic location of each sample were recorded using a GPS device. Sampling was performed at 500 m intervals in two 30-day intervals in cooperation with the Geological Survey of Iran.

About 268 samples were taken for geochemical studies. These samples were taken at distances of 500–800 m from each other following the principles of sediment sampling for geochemical studies, which include sampling from a depth of 25 cm and generally from canal floor sediments.

In addition, for hydrochemical studies and groundwater quality assessment in the study areas, 80 water samples were taken from water wells in the study sites (location of water wells are shown in Figs. 1.3 and 1.4). These samples were taken at random intervals of 20 days from areas where there was access to water wells, covering all parts of the alluvial fan from top to bottom.

1.4.3 Laboratory Studies

This includes laboratory studies related to sedimentology, geochemistry and hydrochemistry, each of which is discussed in the following sub-section.

Fig. 1.3 Laboratory studies of granulation. **a** Separation of gravel and sand samples, (**b** and **c**) Hydrometric studies of fine silt and clay particles from alluvial sediments of Ramhormoz and Dezful regions

1.4.3.1 Sedimentological Studies

At this stage, the hydrometric method or other granulation methods can be used to separate gravel particles from sand using standard sieves, and granulation studies of fine silt and clay particles were also carried out. Folk's (1980) method was used to calculate skewness and mean.

1.4.3.2 Geochemical Studies

For geochemical studies, an XRF device that uses Energy Dispersive X-diffraction (EDX) spectroscopy was used to study oxides and main elements in sediments. To perform mineralogical studies, X-ray Diffractometer (model XRD-7000) was used. To study rare earth elements, the Inductively Coupled Plasma (ICP) method and ICP-MASS device were used by the Geological Survey of Iran.

To study the hydrochemistry of groundwater samples in the study areas, hydrochemical devices such as PH meter, flame photometer (for measuring mass of elements such as sodium, potassium and calcium) and potassium meters were from the laboratory of the Geological Survey of Iran. Using these devices, different hydrochemical properties (such as water acidity, pH; water hardness, TH; electrical conductivity, EC; sulfate, SO4; potassium, K; sodium, Na; and calcium, Ca) of groundwater

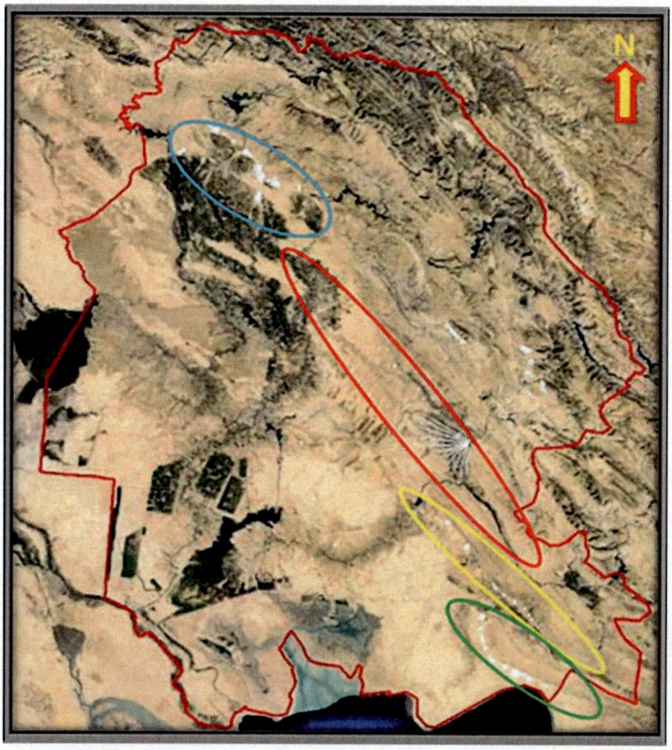

Fig. 1.4 Aerial distribution of alluvial fans in Khuzestan plain (Green: alluvial fans of White Vein; Y: alluvial fans of Omidiyeh; Red: alluvial fans of Shushtar to Ramhormoz; Blue: alluvial fans of Andimeshk and Dezful)

were measured. The duration of geochemical and hydrochemical tests was 6 months from the preparation of the samples to the receipt of the results.

1.4.4 Additional Studies

The final stage of the study includes data collection and data analysis. A detailed review of the results from field studies, granulometry, petrography, geochemistry and hydrochemistry have been compared and analyzed. Various software such as CorelDRAW, SPSS, Adobe Illustrator, ArcGIS, ArcMap, Word and Excel have been used.

1.5 Geographic Location of Alluvial Fans in Southwestern Iran

The distribution of alluvial fans in southwestern Iran (Khuzestan plain from northwest to southeast) can be divided into four parts:

I. White-anticline alluvial fans.
II. Omidiyeh alluvial fans.
III. Alluvial fans from Shushtar to Ramhormoz.
IV. Andimeshk to Shushtar alluvial fans.

White vein anticline alluvial fans are seen in the southeast of Khuzestan along the White Vein fault, and their sediment originates from Aghajari Formation. Omidiyeh alluvial fans start near Mashrageh and continue to Sardasht. These alluvial fans are found along the Aghajari fault, and the source of their sediments are from Gachsaran, Mishan and Aghajari formations. The alluvial fans of Shushtar to Ramhormoz are located along the Ramhormoz fault and their origin is mostly Gachsaran and Mishan formations. Andimeshk to Shushtar alluvial fans are located along Lehbari fault and between Bakhtiari Formation and Khuzestan plain (Fig. 1.6).

In general, the study areas which include alluvial fans of Dezful and Ramhormoz are large alluvial fans that are located in the northern and central parts of Khuzestan province. Ramhormoz alluvial fan is located in the central parts of Khuzestan province, which includes a very large and typical alluvial fan (more than 30 km long and 11 km wide) and is located in the city of Ramhormoz. Dezful alluvial fans also include a large number of small to medium size alluvial fans (5–15 km long and less than 5 km wide) that start from the northwest of Andimeshk and continue to the northeast of Dezful and near Shushtar (Fig. 1.6). It is very difficult to study the origin and changes in sedimentological properties of these alluvial fans because of their small sizes.

References

Bowman B (2019) Principles of alluvial fan morphology. Springer, Netherlands, p 151

Darcy M, Roda Boluda DC, Whittaker AC, Carpineti A (2015) Dating alluvial fan surfaces in Owens Valley, California, using weathering fractures in boulders. Earth Surf Process Landf 40(4):487–501

Dorn RI (2009) The role of climatic change in alluvial fan development. In: Parsons AJ, Abrahams AD (eds) Geomorphology of desert environments. Springer, Dordrecht, pp 723–742

Folk E (1980) Petrography of sedimentary rocks. Hemphill Publishing Company, p 182

Kumar A, Roy SS, Singh CK (2020) Geochemistry and associated human health risk through potential harmful elements (PHEs) in groundwater of the Indus basin, India. Environ Earth Sci 79:86–97

Li K, Deng Q, Hou M (2020) Geochronology and sedimentology of the Huashan Group in the northern Yangtze Block: implications for the initial breakup of the South China. J Earth Sci 167–179

Mazzorana B, Ghiandoni E, Picco L (2020) How do stream processes affect hazard exposure on alluvial fans Insights from an experimental study. J Mt Sci 17:753–772

Miall AD (2014) Fluviall depositional systems. Springer International Publication, p 316

Pourmorad S, Harami R, Solgi A, Ale Ali M (2021) Sedimentological geochemical and hydrogeochemichal studies of alluvial fans for mineral environment purposes. Lithology and Mineral Resources 56:89–112

Pourmorad S, Jahan SH, (2021) A model for comprehensive studies of alluvial fan deposits, Case study: Ramhormoz Mega-fan in southwest Iran. J Earth Sci Climatic Change 12(3)

Rachocki AH (1981) Alluvial fans: an attempt at an empirical approach. Wiely, p 391

Sanchez-Nunez JM, Macías JL, Saucedo R (2015) Geomorphology, internal structure and evolution of alluvial fans at Motozintla, Chiapas, Mexico. Geomorphology 230:1–12

Sissakian VK, Al-Ansari N, Abdullah LH (2020) Neotectonic activity using geomorphological features in the Iraqi Kurdistan Region. Geotech Geol Eng 138–149

Strahler A (1952) Dynamic basis of geomorphology. Geol Soc Am Bull 63:923–938

Tavanaei F, Hassanpour J, Memarian H (2020) The behavior and properties of Tehran alluvial soils under cyclic loading of urban vibrations-a case study: Arash-Esfandiar tunnel. Bull Eng Geol Environ 189–197

Zhang Y, Ye W, Ma C (2020) Middle to Late Holocene changes in climate, hydrology, vegetation and culture on the Hangjiahu Plain, southeast China. J Paleolimnol

Chapter 2
Geology of the Study Area

Abstract It is vital to have sufficient knowledge of the geology of the region, including the type of sediments, sediment origin, formation and morphology of the region, and other geological and environmental information about the region (Pourmorad and Jahan 2021). In terms of sedimentary and tectonic properties, Iran has several basins and sub-basins, of which the most important is the Zagros Basin (Fig. 2.1). The Zagros Basin is characterized by a thick sequence of 7–14 km of sedimentary sediments in large dimensional areas along the north-northeastern edge of the Arabian plate (Bagheri Moghadam and Kharazian 2020). The Zagros Basin has been a stable part of the Gondwana subcontinent during the Paleozoic, and the inactive margin was formed during the Mesozoic (Alavi 2004). The collision between the Iranian and Arabian plates during the Cenozoic led to the formation of the fold-thrust Zagros belt and the associated Foreland Basin (Alizadeh et al. 2020). This folded-trust belt has long been considered due to its huge oil and gas resources and extensive studies have been conducted on it (Pash et al. 2020). Due to the geographical location of the study areas, which is in the folded Zagros, in this chapter we have tried to explain the Zagros tectonics (structural evolution of the Neottis Basin) and the Zagros stratigraphy.

Keywords Zagros basin · Sanandaj-Sirjan region · Folded Zagros · Tertiary stratigraphy

2.1 Structural Divisions of Zagros Basin in Iran

Geographically, the Zagros Basin is in the upstream part of the alluvial fans, which can be divided into three regions: Fars province (central part of Iran), Khuzestan province (southwestern Iran) and Lorestan province (western Iran) (Berberian 1995). The Zagros Basin can be divided into two parts: southeast or Hormoz Basin and northwest or Ahvaz Basin, which is located in the boundary between these two fault basins called Qatar-Kazerun fault (Seraj et al. 2020). Geomorphologically, from northeast to southwest, Zagros includes high Zagros (inner Zagros), folded Zagros and Khuzestan plain (Berberian 1995). According to Alavi (2004), the Zagros orogenic belt in Iran is composed of three tectonic units from northeast to southwest (Fig. 2.2).

S. Pourmorad and A. Mohanty, *Alluvial Fans in Southern Iran*, Advances in Geographical and Environmental Sciences, https://doi.org/10.1007/978-981-19-2045-5_2

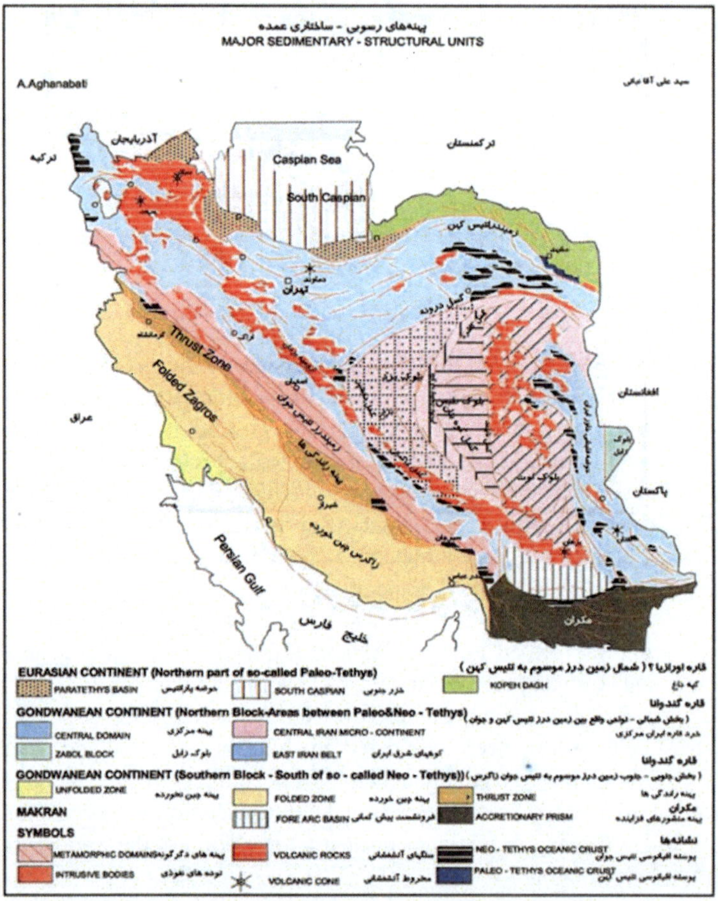

Fig. 2.1 Major sedimentary-structural zones of Iran (Aghanbati 2016)

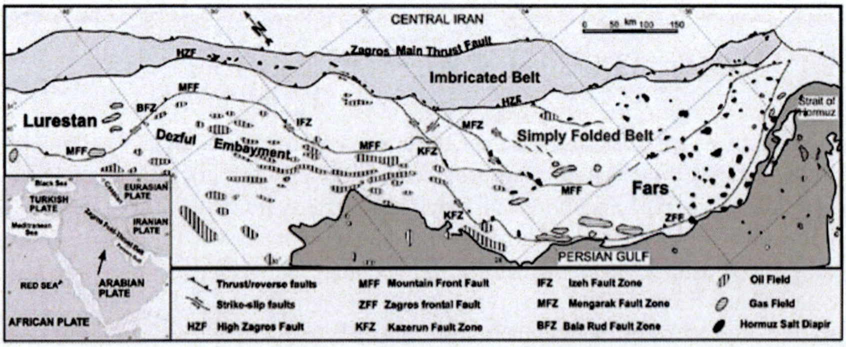

Fig. 2.2 The tectonic framework of the folded belt—Zagros Trust—showing the major fault zones along with the distribution of oil and gas fields (Sepehr and Cosgrove 2004)

- Urmia Magma Collection—Girl;
- Sanandaj-Sirjan region;
- Slightly folded Zagros belt.

2.1.1 Urmia Magma Collection—Girl

This complex is composed of about 4 km thick inner and outer igneous rocks, which are spread almost all along the Zagros. This assemblage is interpreted as the Andean-type magmatic arc (Alavi 2004) and includes various lithological units including gabbro, granodiorite, granite and smaller and larger plutonic sections. In addition, it includes a wide distribution of basaltic lava flows, trachyted basalts, andesite, dacite, trachyte and pyroclastic basalts (Berberian 1995). The Triassic or younger thrust faults are located in the northeastern margin of this complex, and volcanic rocks present there are covered by Tertiary and Quaternary sediments (Alavi 2004). Boge gravity maps show a negative anomaly in this area. The calculation of thickness in this region has shown that on average, the thickness in this region is about 5–10 km more than the average thickness calculated for the Iranian crust (50–45 km). This increase is probably the result of magmatic activity and trust to the northeast (Berberian 1995).

2.1.2 Sanandaj-Sirjan Region

This region is located in the southwest of Urmia Dokhtar magmatic arc in Iran and has an average width of 150–250 km. The northeastern boundary of the Sanandaj-Sirjan region is determined by the lateral location of a number of faults (such as Shirkuh fault; Partabian et al. 2020). The southern boundary of this region does not correspond to the main Zagros fault, but this boundary is defined by slightly curved folds, which are located in front of the thrust sheets, and are miles away from the main Zagros fault (Alavi 2004). The negative gravity anomaly and calculated depth of Moho show an increase in the average crust thickness in this area compared to its average in Iran by 15 km (Sheikholeslami 2015). The only volcanoclasts observed in the Sanandaj-Sirjan region, which include subvolcanic lenses of diabases, thin layers of basalts and rhyodacites, are interbedded with shales in the shallow and coastal areas of Carboniferous, Permian and the Lower Triassic (Alavi 2004). This indicates that the stratigraphic units of Sanandaj-Sirjan, with the exception of the marine sequence to the Middle–Lower Triassic, are mainly volcanic and sediments of inactive and shallow continental margins (Shafaii Moghadam et al. 2013). The middle to Late Triassic is also the time of magmatic activity that exists in the form of gabbro intrusions, granitoids and theolithic lava flows in this zone.

It seems that the transformation of the Middle Triassic to the late and the magmatic activity are related to the expansion processes that led to the formation of the Neuttis

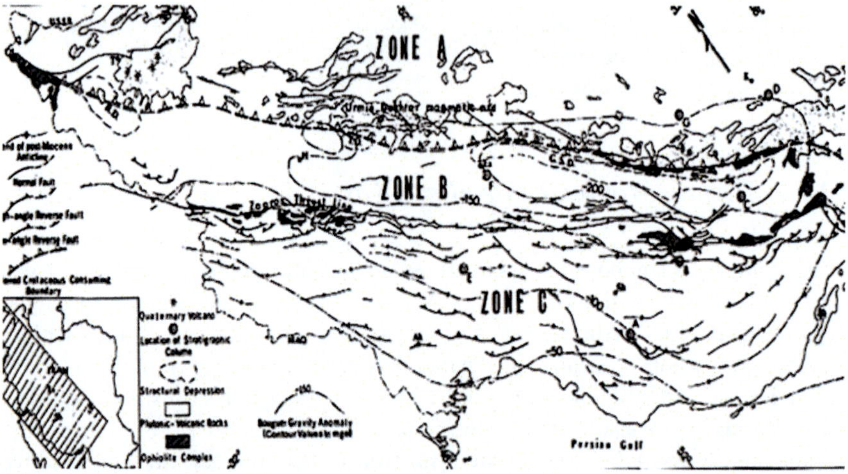

Fig. 2.3 Tectonic division of the Zagros orogenic belt (Alavi 2004). In this figure, zones (C, B and A) represent the Urmia-Dokhtar magmatic assemblage, Sanandaj-Sirjan region and the slightly folded Zagros belt, respectively

ocean. Paleozoic and Mesozoic sequences in this region are medium to low metamorphic rocks. They are non-metamorphic as green schists. It is believed that the metamorphic phase occurred during the Late Cretaceous (Alavi 2004) (Fig. 2.3).

2.1.3 Folded Zagros

The folded Zagros (outer Zagros) has a width of 150–250 km, and forms the marginal and cratonic depression of the Saudi shield (Zaberi et al. 2019). The folded Zagros is continuously subsiding during the Mesozoic and Cenozoic and with the accumulation of thick sequences of sediments (Aghanbati 2016). In the Zagros, the folds of Precambrian to Middle Triassic rocks form the Gondovanai facies, which are similar to other parts of Iran, while the Mesozoic and Cenozoic sequences have rocky facies compared to simultaneous sediments in other parts of Iran, which are also biologically different and mostly represent the southern facies of the young Tethys (Elias et al. 2019). This fact shows that, from the Middle Triassic onwards, the sedimentary conditions prevailing in the folded Zagros have been different from other regions of Iran (Fig. 2.4). Ancient geographical studies show that the folded Zagros does not have the same geological features everywhere (Ehsani and Arian 2015). Figure 2.4 show a comparison between the thicknesses of the stratified Zagros sequences with the high Zagros, which are explained in different sections, summarized below.

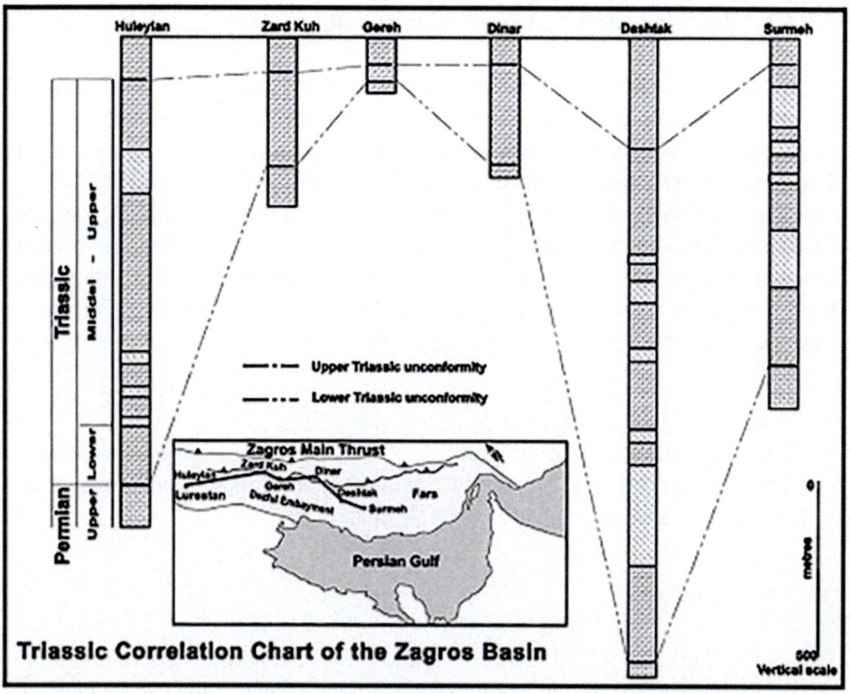

Fig. 2.4 Comparison of the thickness of stratified Zagros Triassic stratigraphic sequences with high Zagros (Szabo and Kheradpir 1978)

2.1.3.1 Lorestan Area

A part of the Zagros is folded, the general trend of which is along the drift zone (high Zagros; Zaberi et al. 2019). The northeastern border is bounded by the Genoa border without drifts, eastern border corresponds to a curvature and western–northwestern border corresponds to the southernmost Zagros anticline, which coincides with the Iran–Iraq border strip. The most important features of the Lorestan region according to Matiei (1993) are

- The northwest-southeast trend has structures consisting of alternating large anticlines (such as Kabir Kooh) and smaller anticlines consisting of the Bangestan Group in the south and Amish Flesh Formation and Garou Formations in the north.
- Large landslides (e.g. Seymareh landslide).
- Gravitational collapses.

2.1.3.2 Fars Area and Post-Arid Bandar Abbas

In general, the Persian Gulf is divided into coastal and inland Fars. Many geologists consider Fars to be located between the Kazerun fault in the west and Minab fault in the east, but Matiei (1993) considered the geological feature of the eastern part of Fars to be different because of the dryness of Bandar Abbas Hinterland.

Thus, the western border of the Fars area is closed to Kazerun fault zone and its eastern border is linear, which continues from Nakhilo port and near Fino Mountain, to north of Abbas port to the main thrust of Zagros. The northern border of the Persian Gulf is the drift zone, and the southern border is the coastline of the Persian Gulf. The most important geological features of the Fars region are

- Having platform conditions due to the continuity of Saudi Arabia, which reaches Persia from Qatar and is referred to as bullfighting.
- Anticlines with different orientations of east–west and even northwest that the change of structures is the result of the operation of piston faults or rotation of the Arabic plate vector relative to the Iranian plate.

2.1.3.3 Izeh Area

Part of the Zagros is folded, which is limited from the north to the southern border of the driving zone, from the south to the northern border of the Dezful depression, from the east to the Kazerun fault and from the west to the hypothetical length of the Balaroud fault (Matiei 1993). One of the characteristics of this zone is that it contains Izeh fault, which is a kind of transverse fault along the right slip and is similar to Kazerun fault. As a result, the Izeh zone is divided into northwestern and southeastern parts (Sarkarinejad et al. 2017). In the northwestern part, the core of the anticlines is composed of Bangestan Group structures and has no oil traps, but in the southeast, the limestone of the Asmari Formation is the core of the anticlines, which shows less uplift and erosion.

2.1.3.4 Abadan Plain

It is a structural zone located at the southwestern end of the Zagros. Its northern and northeastern borders are limited to the Zagros stratigraphic front (southern edge of Susangard anticlines, Ab-e-Timur and Mansouri), and after crossing the south of Rag-e-Sefid Square, it enters the Persian Gulf and its southern border is Abadan plain, Persian Gulf and Saudi Arabia (Aghanbati 2016). Differences in the construction process and deformation style in sedimentary cover are the cause of the separation of Abadan plain and the collapse of Dezful. Pyongyang maps do not show a clear boundary between these two geological zones. The border between these two areas is marked by gentle anticlines that have a northwest-southeast trend (Matiei 1993).

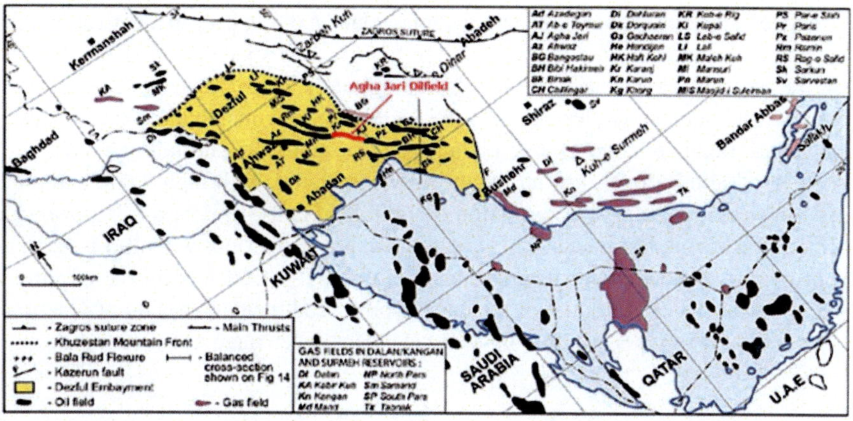

Fig. 2.5 Dezful depression location in the folded Zagros belt (Bahroudi and Koyi 2003)

2.1.3.5 Dezful Fall

Dezful depression is a geological structure in the southwest of the Zagros Trust that covers most of Iran's oil fields (Fig. 2.5). Initially, this was a topographic depression, but in general it refers to an area of the Zagros in which the Asmari Formation has no protrusions (Matiei 1993). The Dezful depression is located between three building phenomena that include the upward bending zone, the mountain front bending zone and the Kazerun bending fault zone. This zone is part of the Zagros Pit and has a slope between 3000 and 6000 m, but it is more stable and less folded in terms of tectonics and folding compared to neighboring areas (Carrubal et al. 2006). The oldest structural evidence in this area belongs to the Upper Cretaceous, but the structures around the Dezful depression and the fractures within it were probably active in the Jurassic and even earlier. These linear structures were still active until the Oligocene or Middle Miocene (Matiei 1993). This area is more stable than its neighboring areas, so it is less wrinkled than Lorestan, Fars and Bandar Abbas districts. According to studies, the shortening of the folded Zagros thrust belt in this area is about 85 km (Pash et al. 2020). Co-thickness maps of Aghajari and Bakhtiari formations clearly show that this area is subsiding due to the thickening of these formations in the area.

2.1.3.6 Main Faults in Zagros Basin

Previous models presented on the expansion of the Zagros shortening only understand how the deformation front migrates to the southern regions in different proportions. In these studies, no attempt was made to explain the facies and sediment thickness at the same time as the tectonic activity in the foreland Zagros basin at the time of shortening and reactivation of piston faults. According to a new tectonic model, during the

Cenozoic, the collision between the Arabian plate and Iran caused reactivation of the piston faults and also shortening of the piston (Sarkarinejad et al. 2017).

Pesang faults are located in two sets of longitudinal and transverse faults along the northwest-southeast (Bahroudi and Koyi 2003). Longitudinal faults with a northwest-southeast trend are almost perpendicular to the direction of shortening with a north-east trend and probably in the form of inverted faults with high angle and slope-slip movements during the formation of the folded belt—the Zagros Trust operated (Fig. 2.6). Pisang transverse faults are divided into two groups with north–south and northwest-southeast trends (Sarkarinejad et al. 2017). Considering the importance of reactivated Pesang faults in the formation of sedimentary basins of Gachsaran Formation, the main Pesang faults that have been reactivated are introduced.

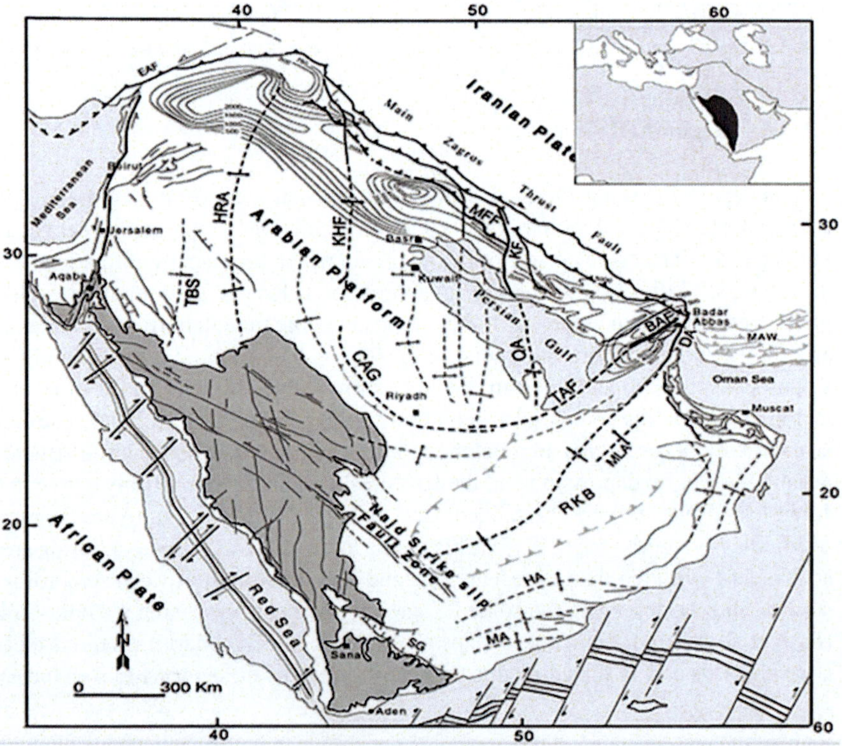

Fig. 2.6 The major faults of Pesang and the linear structures of the Arabian plate including Diba fault (DA), Rab al-Khaimah basin (RKB), Bastan fault (BSF), Bastak fault (BAF), Trans-Arab highway (TAF), Kazerun Fault (KF), Qatar Arc (QA), Mountain Front Fault (MFF), Khaneqin Fault (KHF), Hill Gara Arch (HRA), Tabuk Basin (TSB), Mardin Height (MH), Arabian Central Graben (CAG), ManderLekhwir Elevation (MLA), Route Line (MLA), Hadromat Arc (HA), Mocal Arc (MA), Graben Sadegh (SG), Makran Ascending Wedge (MAW) and Eastern Anatolian Fault (EAF). In this figure, the lines are the same thickness as the Gachsaran Formation in terms of feet (adapted from Bahroudi and Koyi 2003)

Diba Fault

This fault is in the eastern boundary of the Zagros Foreland Basin. It not only separates the Persian Gulf from the Oman Sea at the eastern end of the Strait of Hormuz, but also the eastern boundary of severe and extensive earthquakes in the folded-thrust Zagros belt (Bahroudi and Koyi 2003; Fig. 2.6). Rock facies maps and the thickness of the Zagros Basin have clearly shown that the Diba fault has acted on sedimentation as an important facies divider in the late Mesozoic (Zaberi et al. 2019). The southern regions of Rab al-Khali fault, which is an extended depression along the southeastern desert of Saudi Arabia, can be considered as a southern sequence of Diba fault (Bahroudi and Koyi 2003).

Fault Along the Arabic Landslide—Bastana

This fault is located on the west side of Diba fault with northeast trend, and is composed of two parts, one along the Arab slip and the other along the Bastaneh fault (Matiei 1993). These two are probably connected on Qeshm Island. Bostaneh fault with a length of about 250 km has caused the lateral displacement of the left side of the Zagros anticlines up to distances between 30 and 50 km. Upper Cretaceous–Miocene facies-lithic maps of the Zagros Basin show the effect of this fault on sedimentation in this area (Murris 1980).

Kazerun Fault-Diameter

Among the transverse faults with a north–south trend, the Kazerun-Qatar fault is the most well-known active Pesang fault. This fault has a long history of reactivation along the southern fold-thrust belt of the Zagros, the Persian Gulf and the Arabian Peninsula (Murris 1980). Facies boundaries have shown that the Kazerun-Qatar fault has been repeatedly activated during the expansion of the Zagros Basin. The Qatar-Kazerun fault has limited the Dezful depression to the western regions. It seems that the Kazerun-Qatar fault has caused a lateral displacement of the right side of the Zagros mountain front fault by 140–150 km and 6 km to the west (Aghanbati 2016).

Khaneqin Fault

Part of the border between Iran and Iraq corresponds to the Khaneqin fault. This fault caused the collapse of Dezful from the area behind the mountain (Lorestan province) that separates in the east. This fault has caused a lateral displacement of the right-side fault of Zagros mountain front fault by 130 km. The thick lines of some periods (for example, Jurassic and Upper Triassic) have shown that the sedimentary trends of the region are consistent with the Khaneqin fault (Matiei 1993).

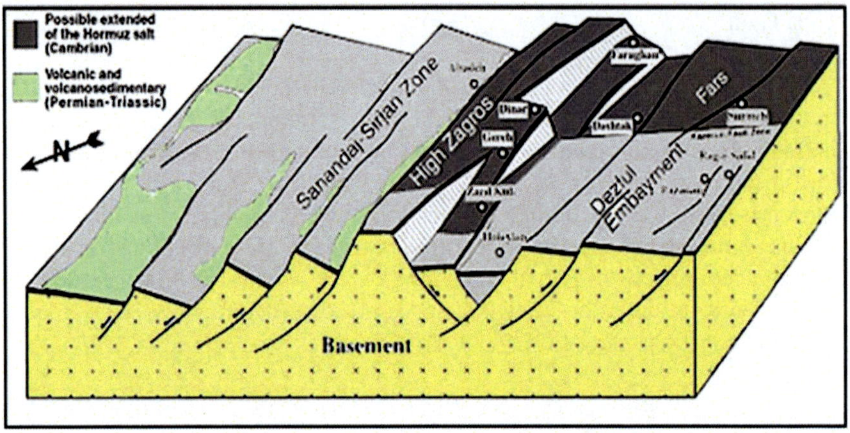

Fig. 2.7 Schematic three-dimensional diagram showing the effect of normal bedrock faults on the sedimentary environment of Permotrias Formations (Bahroudi and Koyi 2003)

Hill Gara Fault

This fault is one of the most important regional trends in the northwestern border of the Foreland Basin of the Zagros (Fig. 2.6). Early to Late Ordovician and Early Silurian rock facies and thick lines have shown that this linear structure has a great effect on the facies and thickness of Phanerozoic sediments (Aghanbati 2016).

It seems that the linear structure of Hill Gara has divided the Iraqi part of the Zagros orogenic belt into two structural blocks, Mosul and Kirkuk. This fault has affected the amount of sediment in these two blocks in such a way that more than 4 km of Phanerozoic sediments have been deposited in Kirkuk block compared to Mosul block (Matiei 1993) (Fig. 2.7).

2.2 Tertiary Stratigraphy of Zagros and the Study Area

The outcrop formations in the study area are mostly Tertiary, which are described in detail here.

In general, Tertiary Zagros sediments can be studied in two parts: Lower Tertiary (Paleocene–Middle Miocene) and Upper Tertiary (Upper Miocene–Pliocene) (Matiei 1993). Two sedimentary cycles can be detected in the Lower Tertiary. These cycles include the Jahrom cycle (Paleocene–Eocene), and the Asmari cycle (Oligocene–Early Miocene). The Upper Tertiary corresponds to the mega-sequence of the 11 folded belts of the Zagros Trust (Alavi 2004), and generally consists of a sequence of seabed sediments that extends from the Upper Miocene to the Pliocene (Fig. 2.8). This section has a summary of the history of existing formations and a detailed study of these formations.

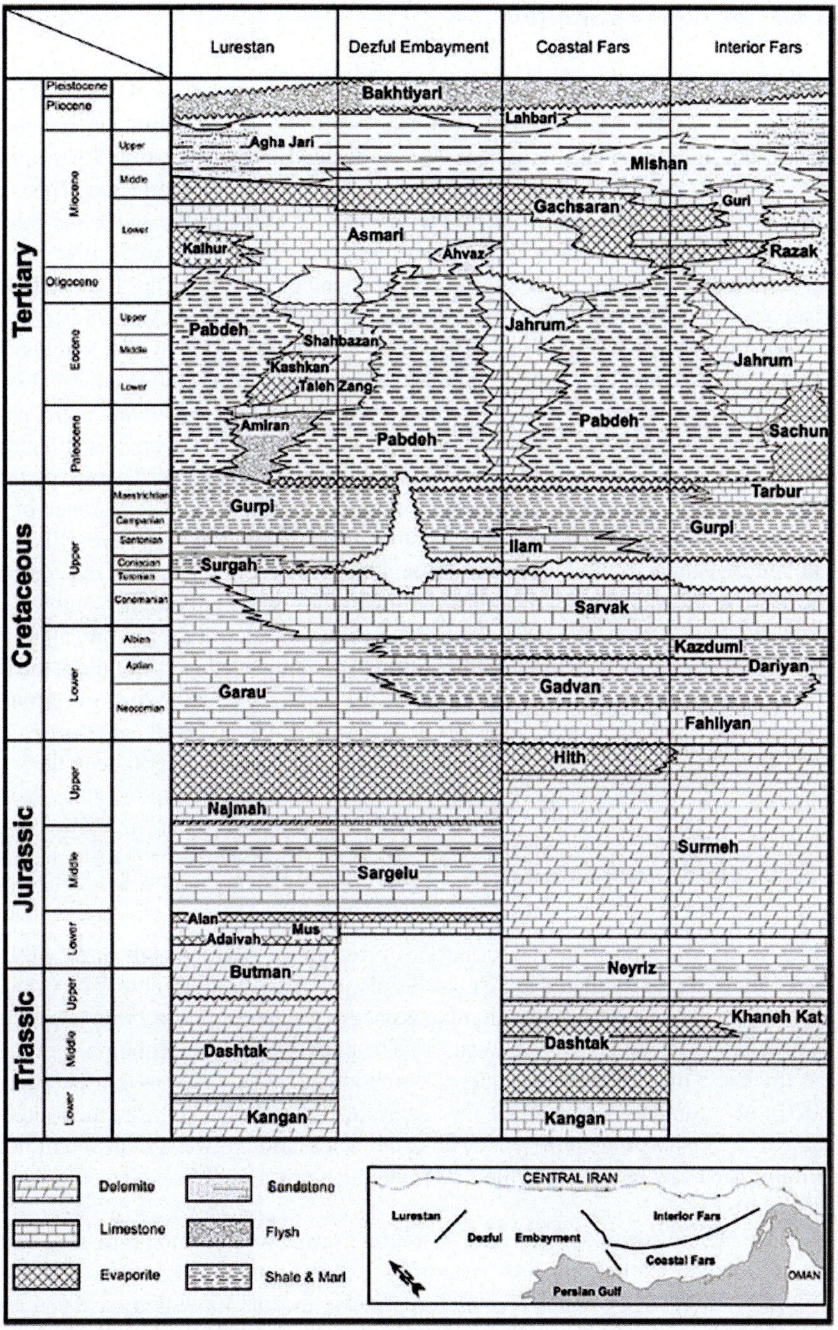

Fig. 2.8 Stratigraphic columns of Zagros in different regions (Sepehr and Cosgrove 2004)

2.2.1 Asmari Formation

Asmari Formation is the youngest oil reservoir rock in the Zagros Basin (Matiei 1993). The name of this formation is adapted from Asmari Mountain (southeast of Masjed Soleiman) and its type section is located in Tang-e Gol-e Torsh in this area (Matiei 1993). The thickness of this formation is 314 m and includes resistant limestones, cream to brown in color. This formation is generally located on the Pabdeh Formation, but in the center of Lorestan, with erosive discontinuity, it is located on the Shahbazan Formation and its upper boundary is formed by the Gachsaran Evaporation Formation. The Asmari Formation has two parts: Ahvaz sandstone and Kalhor evaporation. The pattern cut of Ahvaz sandstone section in well number 6 of Ahvaz and its reference cut in well number one in Timur water field (Matiei 1993). In general, the sandstones of Asmari Formation are divided into lower and upper parts. The lower part sandstones consist of large lenses within the Asmari sedimentary basin but have no outcrop due to their spread on a regional scale. The age of these Late Eocene to Oligocene sediments has been determined. The sandstones of the upper part are a continuation of the sandstone formation of the Iraqi cave, which has entered the Asmari sedimentary basin from the southeastern regions of Iraq and from the north of Kuwait. The evaporative part of Kalhor exists only in the southwest of Lorestan. This section is 118.8 m thick and includes gypsum at the bottom, marl with thin layers of lime in the middle, and gypsum containing two layers of lime at the top. The evaporating part of Kalhor is seen laterally interfering with the Asmari carbonate sequence. A small continuation of this section called Asmari basal anhydrite can be seen along Masjed Soleiman, NaftSefid, Haftgel and Parsi fields (Matiei 1993).

2.2.2 Gachsaran Formation in the Study Area:

Gachsaran Formation is the first formation of Fars group (Mahmoodabadi 2020). This rock formation is considered as Asmari Formation and is equivalent to formations in Iran and Iraq. Gachsaran Formation consists of a sequence of evaporitic rocks such as salt, anhydrite, and red and gray marls. The thickness of this formation varies, but the thickness of its complete sequence sometimes reaches 1600 m (Hashemi et al. 2020). According to Alavi (2004), the Gachsaran Formation is far from the source of the wedge-top accumulation zone. The combined section of wells of different fields is introduced as a model (informal) section which consists of 7 sections (Table 2.1) (Fig. 2.9).

Since none of the fossils of the Gachsaran Formation are among the time index fossils, they do not help much in determining the age of this formation. Therefore, according to the age of the lower and upper classes, this formation has been aged (Monjezi et al. 2019). According to the studies, the base of Gachsaran Formation along the Foreland Zagros Basin is two periods and becomes younger to the northwest with increasing evaporation in this formation (Rahimi et al. 2020). In Qeshm region,

Table 2.1 Pattern cutting (informal) of Gachsaran formation (Matiei 1993)

Thickness (m)	Petrology	Section
137	Alternation of anhydrite, gray marl and limestone (can be divided into 5 zones)	7
278	Anhydrite, red marls and lime (bottom), rock salt (middle) Anhydrite and red marl (top)	6
308	Alternation of anhydrite, red marl, salt rock and thin layers of limestone	5
5/834	Thick rock rotation, gray marls, Anhydrite and quantitative layers of lime	4
225	Anhydrite, thick gray marls	3
5/113	Rock salt, anhydrite, gray marl and thin limestone bands	2
40	5 evaporation cycles including anhydrite, marl, lime and bituminous shale	1

Fig. 2.9 Close view of a part of marl, sandstone, gypsum and limestone rotation in the dominant marl section of Gachsaran Formation (northwest of Ramhormoz-southwest of Iran)

the lower limit of Gachsaran Formation, Pabdeh and Jahrom Formations have been reported (Monjezi et al. 2019). Therefore, the time of sedimentation of Gachsaran Formation in this area is Oligocene (Hashemi et al. 2020). Around the mountain, its age is Early Miocene and along Bushehr (where the second salt basin begins) its age is Early Miocene (Matiei 1993). This formation is also spread in northern Iraq and southeastern Turkey and has a Middle Miocene age in these areas (Fig. 2.10). In summary, it can be concluded that the Gachsaran Formation has been deposited along

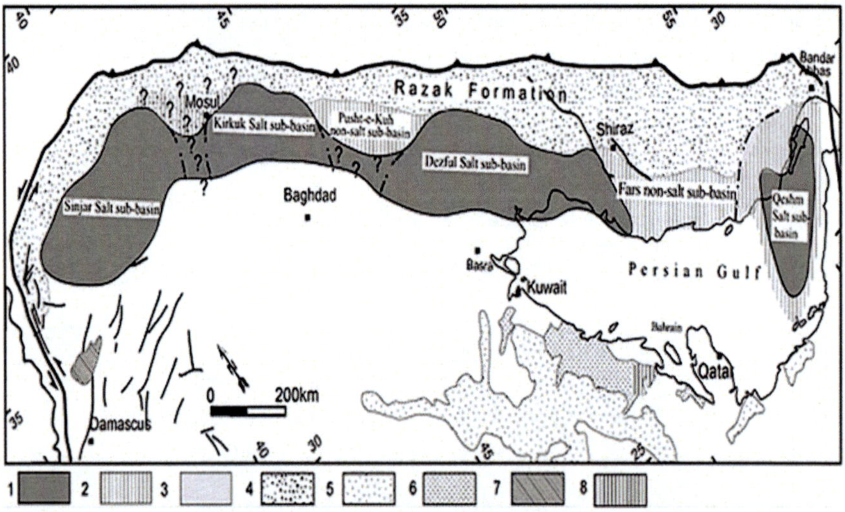

Fig. 2.10 Distribution of facies and temporal equations of Gachsaran Formation along Zagros Basin (adapted from Bahroudi and Koyi 2003). 1—Gachsaran salt facies, 2—Gachsaran non-salt facies, 3—Razak Formation, 4—Gachsaran Formation

the Foreland Zagros Basin for a period of 20 million years (from Oligocene–Middle Miocene) (Rahimi et al. 2020).

This formation is the oldest stratigraphic unit in the study area in Ramhormoz city, which has the largest outcrop in the Persian anticline in the northeast of the region, and is located on Aghajari and Mishan formations due to Ramhormoz drift (Fig. 2.11). This unit is mostly composed of evaporitic facies such as gypsum and anhydrite with alternations of greenish to reddish-gray marls, between which thin to medium layers of limestone and sandstone can be seen. Parasitic folds are also seen in this formation. This formation generally consists of two parts, gypsum and greenish-gray marl in the lower part, and red marl and gypsum in the upper part. In the lower part (greenish-gray gypsum and marl), the alternation of medium, thick to very thick layers of gypsum and anhydrite with gray to greenish-gray marls is alternated (Fig. 2.12a).

In some places, the nodular structure is also found within the marls (Fig. 2.12b). The predominant lithology of gypsum in this unit has increased the layering disorder, and parasitic folds become visible. The upper part of the Gachsaran Formation consists of a period of red marls, and the middle layers are thick to very thick gypsum (Fig. 2.12c). Gray marl layers are also seen in small numbers in this unit, especially in the vicinity of gypsum layers. Thin to medium layers of limestone sandstone and sandstone limestone with cubic fractures are seen in the highest part of Gachsaran Formation (Fig. 2.12d). The predominant marl lithology and red color are the features of this section. There is less clutter in this unit, and it has more regular layering than the lower parts. The lower contact of this formation is not seen in the study area, but outside the study area with Asmari Formation it has an erosive discontinuity. Its

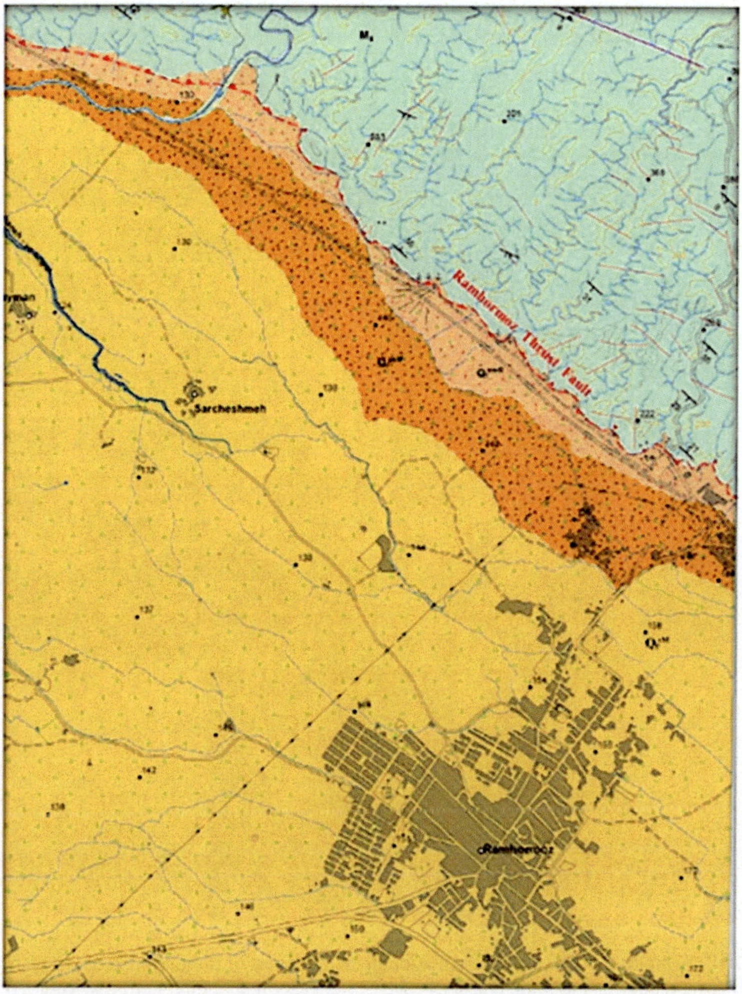

Fig. 2.11 Location of Gachsaran Formation in Ramhormoz area (green) in relation to Ramhormoz city (the center of Ramhormoz large alluvial fan) and Quaternary sediments (dark and light brown). Gachsaran Formation is located on Aghajari and Mishan Formations due to Ramhormoz drift and as a result this formation is observed in the vicinity of Quaternary (Pourmorad and Jahan 2021)

upper contact outside the area has alternating marl and Gachsaran limestone to marl and limestone of the formation Mishan, and the slope is steep and gradual. Due to the surface protrusion of gypsum in this formation and re-dehydration of anhydrite, often secondary gypsum in the form of alabaster, enterolithic texture is seen, which is caused by contact with groundwater in arid rocks (Tucker 1981; Fig. 2.12f). Other constructions in gypsum include poultry netting, porphyrotopic texture, satin spar and selenite gypsum in cavities (Fig. 2.12g).

Fig. 2.12 Lithology of Gachsaran Formation in the northeast of Ramhormoz. **a** Red marl, gypsum and gray marl alternation in the northern edge of the Persian anticline (northeast view). **b** Nodular texture in Gachsaran Formation. **c** Red gypsum and marl rotation of the upper part of Gachsaran Formation (northeast view). **d** Calcareous sandstones with a cubic fracture in Gachsaran Formation (looking to the northwest). **e** Intra-rock texture in gypsum in Gachsaran Formation. **f** View of secondary alabaster gypsum in Gachsaran Formation. **g** Construction of poultry net in rock Gypsum Formation in Gachsaran Formation

2.2.3 Mishan Formation

This formation has no outcrop in Ramhormoz and Dezful areas, but has been described as a part of Tertiary stratigraphy in Zagros. The name of this formation is taken from Mishan village located 50 km southeast of Gachsaran in Kohkiluyeh province, and its pattern section is introduced along the road that passes through the southern edge of Gachsaran oil field (Fig. 2.13). Toward the end of Bordigalin, the

Fig. 2.13 Pathways of Mishan Formation in Baghmalek region of Khuzestan (Pourmorad 2018). **a** Fault boundary and expulsion of Asmari Formation on Mishan Formation (northeast view). **b** Alternation of red marl and gray sandstone to the middle cream layer of Mishan Formation in the southwest of Barangerd village (northwest of Ramhormoz area). **c** Periodically from gray marl, thin to medium layer cream sandstone and between layers of thin layer limestone in the northwest of Baghmalek (northwest view)

subsidence of the area is located between the Persian platform and the collapse of Dezful, where the marine environment has expanded, and the Gachsaran Formation has been covered with a progressive and shallow sea.

The pattern section of Mishan Formation (in Gachsaran oil field) includes 710 m of gray marl and clay limestones consisting of fossil shells and other shell fragments (Matiei 1993). The lower part of the Mishan Formation (~60 m thick) is mostly off-white limestone, named as worm-layered layers, and to the southeast it is replaced by reef limestone in the Guri section of the Mishan Formation. In previous studies, this part was called Gori Formation or Apercolina Lime (Matiei 1993).

2.2.4 Aghajari Construction in the Study Area

The name of this formation is adapted from Aghajari city in southwestern Iran (Aghanbati 2016). In this section, Aghajari Formation is 2966 m thick, with repetitive rotation of cycles that are granulated upwards, and is located on a slope below Bakhtiari Formation (Fig. 2.14). In Dezful depression, this formation has the highest thickness, but to the east and southwest, the thickness of this formation decreases. The

Fig. 2.14 Location of Aghajari Formation: (Aj) below Bakhtiari Formation, and (Bk) with discontinuity in northeast of Andimeshk

age of Aghajari Formation is from the Late Miocene to Pliocene based on the age of the fossils present in this formation (Matiei 1993). The predominant lithology of this formation with a thickness of about 1350 m includes a period of thick red marl and medium to thick layer sandstones, cream to gray color, with layers from mudstone to silt. Medium to thin red sandstone is also found in this section (Aghanbati 2016).

In the study area, the dominant lithology of this formation consists of a period of thick red marl and medium to thick layer sandstones, cream to gray in color, and layers from mudstone to siltstone. A thin layer of red sandstone is found in this collection (Fig. 2.15). Sedimentary sequences of this formation have a finite upward trend and show different sedimentary structures in different parts of the sequence (Fig. 2.15), and a change in grain size, depth and viscosity. The flow and geometry of waterways are at different times (Jalilian 2018). The largest volume of sediments in Aghajari Formation are red marls and lichens. These sediments are related to the sedimentation of the suspended load of the river in the flood section (Miall 2014),

Fig. 2.15 From the rotation of red marl and thin to thick sandstones (view from northeast of Andimeshk)

which sometimes exists between the layers of fine-grained sandstone with medium to thin layering, which is related to the formation of wide crosses in this area (Obeid et al. 2016; Fig. 2.16).

The rock layers of this formation are exposed in the northeast of Dezful region and have formed moderate ridges in which many joints are perpendicular to each other (Figs. 2.7 and 2.18). This formation is not observed in Ramhormoz region due to being covered by Gachsaran Formation because of Ramhormoz drift. Sedimentary sequences of this formation in Dezful and Andimeshk regions have a finite upward trend with different sedimentary structures that can be seen, which show the change in grain size, depth and viscosity of flow and waterway geometry, formed at different times (Jalilian 2018) (Fig. 2.17).

Fig. 2.16 Massive red marls with intermediate to thin sandstone layers. This part of the facies belongs to the flood plain and the edge of the canal (View from northeast of Dezful)

Fig. 2.17 Face-forming sandstone ridges alternating with marl

Fig. 2.18 Perpendicular joints in the middle of the sandstone layers. The southern edge of Dali Mountain in the north of Andimeshk

2.2.5 *Lahbari Member in the Study Area*

The pattern cut of this member is measured in Tang-e-Takab, located 10 km northeast of Haftgel city, which includes 1575 m of siltstone, silt-gypsum marl, carbonate sandstone and gypsum, where one of its features is large. The seeds grow upwards. This member is pea to earth in color, which is difficult to separate from the younger units resulting from the erosion of Aghajari Formation and Bakhtiari Conglomerate. The study on Hipparion horse belongs to Pliocene, among the works found in Lehbari (Matiei 1993).

This member is observed in the study areas only in the northwest of Andimeshk region and is considered as one of the sources of alluvial deposits of alluvial fans in this region. The lithology of this member consists of Aghajari Formation with a thickness of about 800–1350 m, mainly red marl and siltstone marls and siltstone, which alternate with sandstone, conglomerate and mudstone, and is medium to thick layered (Fig. 2.19). The color of the sandstone and mud layers is cream to gray and the conglomerate layers are gray. The general appearance of this formation is dilapidated and often hilly, and in places where the thickness of hard layers such as sandstone or conglomerate increases, it has formed not very high ridges (Fig. 2.19a). The highest extent of this member is in the west and northwest of Andimeshk and from the west of Andimeshk to the northern parts of the region, the frequency between conglomerate layers decreases and the thickness between sandstone layers increases. In some parts, with the complete predominance of marl and mud layers, gypsum laminates in the form of selenite and satin spars, and some unripe brown coals are also observed (Fig. 2.19b). This member, like Aghajari Formation, is not visible in Ramhormoz region due to the drift of Gachsaran Formation.

Fig. 2.19 The appearance of Lehbari member in Andimeshk region. **a** The appearance of a flat and soft hill, with medium and relatively steep layers of conglomerate—toward the northeast of Andimeshk

2.2.6 Bakhtiari Formation in the Study Area

Bakhtiari Conglomerate Formation is characterized by alluvial-foothill sediments resulting from elevation erosion, which mostly consists of conglomerate and calcareous sandstone that is sometimes deposited on high slopes and sometimes steeply on older formations. The pattern section of this formation in the north of Masjed Soleiman (Godar Lander) includes 550 m of conglomerate with parts in the boulder area, of various ages, which are cemented with coarse calcite and clay. The thickness of this formation is different in different regions, so the Bandar Abbas area is 1027 m and Behbahan (well number 11) is 1330 m (Matiei 1993). The lithology of this formation is about 740 m thick, consisting of medium to very thick layers of conglomerate, gray to cream in color.

This formation is observed in Dezful study area in the northern part of Andimeshk and Dezful cities and borders with Aghajari Formations and Lahbari section and is considered as the main source of sediment supply in these areas. In Ramhormoz region, this formation has no outcrop, and the sediment is deposited through rivers such as Mal Agha from Baghmalek and Izeh counties, which are considered as the source of sediments in these areas. The lower boundary of this formation in Andimeshk region is gradually located on Lahbari member everywhere (Fig. 2.20a). This formation in the study area mainly has horizontal and lens-shaped layers (Fig. 2.20b).

In the study area, the conglomerate layers are cream to light brown in color and thick to layered maroons and greenish-gray to pea-colored. The layers of the conglomerate are composed of calcareous gravelly grains and to a lesser extent grazing. The space between the layers is filled with carbonate cement. The roundness of the rubble is good but the sorting is poor. Calcareous gravels are larger and have a better sphericity compared to snails. Gradual and cross-stratified grain structures are seen within the conglomerate (Fig. 2.21).

Fig. 2.20 **a** The boundary of Bakhtiari Formation with Lahbari member in a deformed form in the northwest of Andimeshk (northwest view). **b** Existence of horizontal and lens-shaped layering in gravel channels of bed load sediments in Bakhtiari conglomerate of Dezful and Andimeshk regions (northwest view)

Fig. 2.21 View of Bakhtiari Formation (as a source rock) in Andimeshk region. **a** Northeast of Andimeshk, and **b** North of Shushtar

The cement and binding materials are mostly calcareous, and snooze materials are scattered with the grains (Fig. 2.22a). The conglomerate layers of this section are thick to medium, convex–concave shaped and jagged contacts due to their high compaction. One of the salient features of these layers in terms of geomorphology is the production of large rock blocks, which have created many debris on steep slopes (Fig. 2.22b). As mentioned, the most abundant sedimentary structures observed in these layers are horizontal stratification and embossing (Fig. 2.22c). The presence of parts with good roundness and medium sphericity is a sign of transportation and high current strength (Jalilian 2018). The support grain texture and the presence of the impregnation fabric indicate the deposition of these sediments by the tensile currents and the bed load of the cut rivers and alluvial fans (Miall 2014). Petrographically, these rocks are composed of relatively rounded limestone fragments that are present among the limestone cement particles. There are snooze and radiolarity pieces in the space between the grains, and a few sand-sized quartz have also been observed in this collection.

Fig. 2.22 Lithological properties of Bakhtiari Formation in Andimeshk. **a** Existence of red nap pieces in the conglomerate layer. **b** Large stone blocks from the conglomerate layers located in the northeast of Andimeshk. **c** Close view of the conglomerate stone blocks in the northwest of Andimeshk

2.2.7 Quaternary Sediments in the Study Area

These sediments are namely a collection of tangled and mostly destructive sediments (debris) that results from the destructive and erosive activities, and are transported to the depo-center by transfer agents such as water and wind. All the downstream and sloping areas, including the bed of canals and flat lands, are covered by these sediments. Lithologically, the composition of this complex is different and is mostly a function of lithology and the type of rock units that are exposed upstream, and it includes sedimentary, alluvium and alluvial sediments that are due to the destruction of sedimentary rocks in Gachsaran, Aghajari and Bakhtiari formations in the basin. The sediments have been released in the form of rock and gravel fragments and have gradually accumulated in low-slope and downstream areas of the slopes and heights. Another group of materials forms alluvial sediments, as a result of the destruction of existing formations that have been transported to downstream areas by running water or during floods.

These sediments in Ramhormoz region originated mainly from rivers named 'Ala and Khanami'. These are the cut rivers, having a length of 18 km and flowing from northwest to southeast from the heights of Gachsaran Formation toward the flood plain. The sediments of the bed of these rivers are mainly in the parts close to the

Fig. 2.23 View of alluvial sediments in the study area—northeast of Ramhormoz

altitudes, including coarse-grained parts with medium sorting and rounding, which increase their circulation as they move away from the top of this river.

In the northern parts of Andimeshk, alluvial sediments are mainly caused by floods, which are coarse-grained and cover the bed of the canal. A significant part of these sediments is the result of destruction of Bakhtiari Conglomerate Formation. In flooding conditions, different sediments move at high speed from the heights of the mentioned areas to downstream of the basin, as well as while passing through low-slope areas due to energy loss, depending on their grain sizes. The sediments deposited in these areas formed the cone shape of the existing alluvial fans. The volume and size of the detrital material decrease from the top of the cone to the bottom. The suspended load of runoff has also deposited over long distances and provided lands with suitable soil texture for agriculture (Fig. 2.23).

In general, Quaternary sediments are relatively widespread in the study areas and include sediments of alluvial fans, alluvial barracks, riverbed sediments and slope debris that are observed in the slopes of the formations in the study region. Except for alluvial fan deposits, which are the main subject of this research, the following section describes other Quaternary sections in the study areas.

2.2.7.1 Fluvial Terrace

Alluvial or river barracks are geomorphological complications caused by rivers (Sugai 2016). In a valley, due to the lowering of the river basin level in its sediments, the remains of alluvial sediments overlook the current bottom of the river, thus creating an alluvial garrison and the river flows at a lower level (Pourmorad and Jahan 2021).

Fig. 2.24 A view of the expansion of alluvial barracks in the Ramhormoz area along the Khanami River (southwest view—Primary Source)

According to previous studies conducted in Ramhormoz region, three types of alluvial barracks with the names Qt2, Qt3 and Qal can be identified, which in Andimeshk region are generally recognizable as Qal section.

Qt2 Alluvial Terrace

This unit is one of the complications that are observed dominantly at the study site and is considered as one of the main factors of sediment transport. The rapid erosion of the deposits of the Fars group, especially the Gachsaran Formation, which has increased as a result of the Ramhormoz drift, has caused the accumulation of these deposits.

In Dezful region, this part consists of silt and clay along with some fine-grained sand in the form of a mass. Gravel fragments are also observed in the form of a dominant matrix that gradually transform into dominant grain gravel fragments. These deposits are composed of fine-grained upward sedimentary sequences, each of which begins with a conglomerate horizon at the base and ends in mud (silty-clay) deposits at the top.

In Ramhormoz region, these deposits are sometimes seen in the form of conglomerate lenses. The upper fine-grained sections in Ramhormoz region are sometimes gypsum due to the abundance of gypsum crystals. Erosion of these deposits has caused the appearance of satellite hills in some parts. The height of these barracks reaches more than 3 m above the base level of the Shifa and Khani rivers (Fig. 2.24).

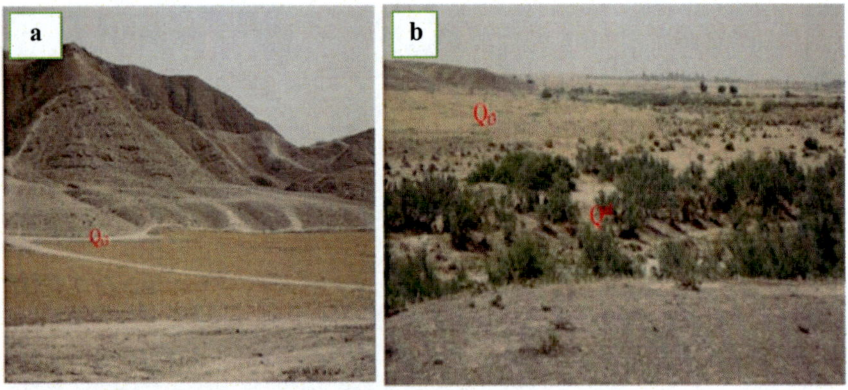

Fig. 2.25 Outline of Qt3 alluvial units in the studied areas. **a** In the middle of Gachsaran Formation (view from northeast in Ramhormoz region). **b** Outcrop of Qt3 alluvial units in Andimeshk region (view from northeast—Primary Source)

Alluvial Unit Qt3

This unit has deposits of young alluvial barracks found around most of the waterways and rivers in the study regions. It is mostly composed of fine silty-clay deposits with sand and gravel lenses, especially in its lower parts. Its height is ~2–3 m from the base level of waterways, which has created suitable areas for agricultural activities (Fig. 2.25a, b).

Qal Alluvial Unit

It contains loose and isolated deposits in the bed of the Madani and Shifa rivers, studied rivers in Ramhormoz region. In Andimeshk region, it can be seen in the bed of seasonal rivers. They are composed of silt clay, and the grains are mostly carbonate, sandstone and to a lesser extent in Ramhormoz region gypsum crystals that have originated from Fars group. These deposits become finer in grain from the heights to the plains of Ramhormoz and Dezful (Fig. 2.26).

2.2.7.2 Deposits of Alluvial Plain

These deposits, which can be seen in the southern part of Ramhormoz, and the western part of Dezful, include fine-grained silty-clay and sandy sediments that are in the continuation of alluvial fan deposits, which results from the gradual conversion of alluvial fans. These deposits are finer in the plain than the alluvial fan, which consists of lenses and horizons of coarse-grained sandy deposits.

Fig. 2.26 Qal alluvial unit deposits in the studied area. **a** Qal unit deposits along the Shifa River toward the north. **b** Qal deposits near the city of Andimeshk—toward the northwest of Andimeshk (Primary Source)

This unit has a relatively smooth appearance and a very mild topographic slope in both study areas, which has been used as agricultural land in many areas, especially in the Andimeshk region.

In some places, especially in the vicinity of large waterways, these deposits have a worn appearance and a mound due to erosion (Fig. 2.27).

Fig. 2.27 Sediments of alluvial plain in Ramhormoz plain—southwest of Darvishan village (view toward the northeast)

Fig. 2.28 Regulations of Ramhormoz region. **a** Northeast of Shifeh village (look to the east). **b** In the middle of Gachsaran Formation (Primary Source)

2.2.7.3 Regulite Soils in Ramhormoz Region

It contains alluvial deposits, including hard and non-hard rock materials that cover the surface of the bedrock. This unit, which is mostly visible in Ramhormoz area, is the result of bedrock erosion, and is often surrounded by rock outcrops. These deposits are often transported, which make their grains somewhat angular. The color of these sediments is affected by bedrock, and is often the color of marl of the Gachsaran Formation (Fig. 2.28).

References

Aghanbati A (2016) Geology of Iran. Geological Survey and Mineral Exploration Organization, Tehran, 1st edn, 708 pp

Alavi M (2004) Regional stratigraphy of the Zagros Fold-Thrust Belt of Iran and its Proforeland evolution. Am J Sci 304:73–97

Alizadeh A, Hormozi H, Moghadam M, Seraj M (2020) DEM-derived geomorphic indices for assessment of tectonic activity at the Dara anticlinal oil structure within the Zagros fold-thrust belt, southwestern Iran. Arab J Geosci 13:192–212

Bagheri Moghadam H, Kharazian N (2020) Morphologic and chemotaxonomic studies of some Teucrium L. (Lamiaceae) in Zagros region, Iran. Iran J Sci Technol Trans Sci. Online

Bahroudi A, Koyi HA (2003) Tectono-sedimentary framework in the Zagros foreland basin. Mar Pet Geol 33:1–16

Berberian M (1995) Master "blind" thrust faults hidden under the Zagros folds: active basement tectonics and surface morphotectonics. Tectonophysics 241:193–224

Carrubal S, Perottil CR, Buonaguro R, Calabro R, Carpi R, Naini M (2006) Structural pattern of the Zagros fold-and-thrust belt in the Dezful Embayment (SW Iran). Geological Society of America, pp 69–87

Ehsani J, Arian M (2015) Quantitative analysis of relative tectonic activity in the Jarahi-Hendijan basin area, Zagros, Iran. Geosci J 19(4):751–765

Elias Z, Sissakian VK, Al-Ansari N (2019) Assessment of the tectonic activity in northwestern part of the Zagros Mountains, Northeastern Iraq by using geomorphic indices. Geotech Geol Eng 1–13. Hashemi et al (2020)

Jalilian A (2018) Advanced sedimentary environments. Payame Noor University Press, 347 pp

Mahmoodabadi RM (2020) Facies analysis, sedimentary environments and correlative sequence stratigraphy of Gachsaran formation in SW Iran. Carbonates and Evaporates 35:25–39

Matiei H (1993) Geology of Iran, Zagros Stratigraphy. Geological Survey of Iran, 536 pages

Miall AD (2014) Fluviall depositional systems. Springer International Publication, 316 pp

Monjezi N, Amirshahkarami M, Bakhtiar HA (2019) Palaeoecology and microfacies correlation analysis of the Oligocene-Miocene Asmari formation, in the Gachsaran oil field, Dezful Embayment, Zagros Basin, Southwest Iran. Carbonates and Evaporates 34:1551–1568

Murris RJ (1980) Middle East, stratigraphic evolution and oil habitat. Am Asso Petrol Geol Bull 64:598–617

Obeid MA, Khalil AES, Azer MK (2016) Mineralogy, geochemistry and geotectonic significance of the Neoproterozoic ophiolite of Wadi Arais area, south Eastern Desert, Egypt. Int Geol Rev 58:687–702

Partabian A, Bagheri S, Morshedi F (2020) Documentation of the SirjanOrocline in the southeast Sanandaj-Sirjan Zone, Iran. J Earth Sci 108:365–389

Pash RR, Sarkarinejad K, Ghoochaninejad HZ (2020) Accommodation of the different structural styles in the foreland fold-and-thrust belts: northern Dezful Embayment in the Zagros belt, Iran. Int J Earth Sci (GeolRundsch) 109:959–970

Pourmorad S (2018) Sedimentary mineral resources exploration. Danaeshyaran Publication, Iran, Teharn, p 385

Pourmorad S, Jahan SH (2021) A model for comprehensive studies of alluvial fandeposits, Case study: Ramhormoz Mega-fan in southwest Iran. J EarthSci Clim Change 12(3)

Rahimi MR, Mohammadi SD, Beydokhti AT (2020) Effects of mineral composition and texture on durability of sulfate rocks in Gachsaran formation, Iran. Geotech Geol Eng 38:2619–2637

Sarkarinejad K, Zafarmand B, Oveisi B (2017) Evolution of the stress fields in the Zagros Foreland Folded Belt using focal mechanisms and kinematic analyses: the case of the Fars salient, Iran. Int J Earth Sci 107:611–619

Sepehr M, Cosgrove JW (2004) Structural framework of the Zagros Fold–Thrust Belt, Iran. Mar Pet Geol 21:829–834

Seraj M, Faghih A, Motamedi H (2020) Major Tectonic lineaments influencing the oilfields of the Zagros Fold-Thrust Belt, SW Iran: insights from integration of surface and subsurface data. J Earth Sci 31:596–610

Shafaii Moghadam H, Robert JS, Rahgoshay M (2013) The Dehshirophiolite (central Iran): geochemical constraints on the origin and evolution of the Inner Zagros ophiolite belt. Geol Society of America Bulletin 75083-0688

Sheikholeslami MR (2015) Deformations of Palaeozoic and Mesozoic rocks in southern Sirjan, Sanandaj-Sirjan Zone, Iran. J Asian Earth Sci 106:130–149

Szabo F, Kheradpir A (1978) Permian and Triassic stratigraphy, Zagros basin, south – west Iran. J Pet Geol 1:57–82

Tuker ME (2001) Sedimentary Petrology: an introduction to the origion of sedimentary rocks. Blackwell, Scientific Publication, London, p 260

Zaberi M, Grutzner C, Navabpour P, Ustaszewski K (2019) Relative timing of uplift along the Zagros Mountain Front Flexure (Kurdistan Region of Iraq): constrained by geomorphic indices and landscape evolution modeling. Solid Earth 10:663–682

Chapter 3
Advanced Sedimentology Studies

Abstract Sedimentary rocks have precipitated in the past period of geology in different natural environments that are nowadays (Pourmorad 2018). The study of the terms and their sediments and processes helps to understand more of their old equivalent (Sarraj and Mohialdeen 2020). Environment and sedimentary processes, old geography and old weather can all be deduced from the study of sedimentary rocks (Li et al. 2020). In this chapter, it is attempting to provide accurate recognition of all sediment properties of these sediment properties, grain size studies, petrographic studies and facies studies, according to extensive studies in studies.

Keywords Sedimentary facies · Granometric · Hydrological parameters · Classification of alluvial fans

3.1 Granometric

Granulometry (granulometry) is the measurement of particle diameter and density ratio of grains. In this study, samples are screened after preparation by al-Locks (Hernandez-Hinojosa et al. 2019). Particle size analysis can be used to determine the sedimentary environment and identify sediment and flow processes (Morales 2019). In studies in the southwest of Iran, in order to accurately study the studies that are based on all sediment studies, samples prepared from the listed areas were analyzed (Table 3.1). In this section, the results of the analysis of the grains and the statistical data are discussed.

3.1.1 Histogram Curves and Normal Distribution

Histogram is a stretching method, with the relative frequency of particle size by different regions (Roy and Banerjee 2002). The normal distribution curve is also a stretching method that is obtained from the binding of the data points. In this study, the histogram and normal distribution curves for each of these sections are separately mapped and examined. For this purpose, 6 samples from different departments of

S. Pourmorad and A. Mohanty, *Alluvial Fans in Southern Iran*, Advances in Geographical and Environmental Sciences, https://doi.org/10.1007/978-981-19-2045-5_3

Table 3.1 Results of some samples in Dezful, southwest of Iran (primary source)

D8	D7	D6	D5	D4	D3	D2	D1	Sample number
32° 26′ 57.4″–48° 27′ 36.70″ E	32° 27′ 2.24″–48° 27′ 54.27″	32° 27′ 30.11″–48° 28′ 27.18″	32° 27′ 30.11″–48° 28′ 27.18″	32° 27′ 22.41″–48° 29′ 4.19″	32° 27′ 24.31″–48° 29′ 30.19″	32° 27′ 48.38″–48° 29′ 54.49″	32° 28′ 34.17″–48° 30′ 7.35″	Location
6.21	6.49	7.07	6.36	7.13	8.65	9.05	9.14	−5
9.77	8.02	9.14	10.12	9.96	7.16	8.25	8.13	−4
10.29	10.19	10.12	10.42	10.54	10.14	11.02	10.29	−3
13.19	14.88	14.47	15.19	14.32	14.11	15.82	15.19	−2
12.34	12.74	13.09	13	12.34	12.09	11.98	12.86	−1
10.11	9.14	7.16	7.12	7.07	6.34	6.47	8.34	0
9.14	8.57	8.02	9.14	9.01	8.43	7.37	8.27	1
8.07	7.19	7.26	6.02	7.83	7.54	6.31	6.34	2
9.11	9.11	9.09	8.04	8.19	8.89	7.34	7.16	3
5.19	5.62	5.83	6.07	6.19	6.16	6.31	6.21	4
6.58	8.05	8.75	8.52	7.42	10.49	10.08	8.07	<4
51.8	52.32	53.89	55.09	54.29	52.15	56.12	55.61	Gravel %
41.62	39.63	37.36	36.39	38.29	37.36	33.8	36.32	Sand %
6.58	8.05	8.75	8.52	8.42	10.49	10.08	8.07	Mud %

Dezful (2 samples from the vertex, Two samples from the middle part and 2 samples from the end of the sediments of the conjugated deposits of the Dezful region were harvested from the canal floor) and 6 samples from Ramhormoz conjugate (2 samples from the vertex, two samples from the middle part and 2 samples from the end of the conjugate of Ramhammed were harvested from the canal floor) were studied (Figs. 3.2 and 3.3).

The results of histogram diagrams and normal distribution of the 6 samples examined in different parts of Dezful conjugates show that the samples studied in the upstream area (D5, D8 samples) belong to particles with size −2 to −4 fi (moderate to coarse grains). The average of gravel deposits studied in the upstream area is 53.4%, the average of sandy sediments is 39% and the average of glyced sediments is 7.6%. In the samples studied in the middle part of the Dezful region (D15, D19 samples), the particles mostly belong to particle size of 1 to −1 (very coarse sand).

The average of gravel sediments is 39.2%, the average of sandy sediments is 44% and the average of glyced sediments is 16.8%. The samples studied in the end departments of the conjugate (D23, D26 specimens) show that the most abundant particles belong to particle size of 4 fetal and clay. The mean of gravel sediments of these samples is 18%, the mean of sandy sediments is 34.4% and the mean of glyced sediments is 47.6% (Fig. 3.1).

The results of the study of histogram diagrams and normal distribution in the studied samples in different parts of Ramhormoz indicate that in the samples studied in the high sections of this conjugal (R2, R6 specimens), The most abundant percentage of particles belonging to particles-3 to -4 fi (moderate to coarse gravel). The mean of gravel sediments of these samples is 31.2%, the mean of sandy sediments is 44% and the mean of glycase sediments is 24.8%. The study of the samples of the intermediate section of the studied conjugate (R10, R13) shows that the most abundant particles of these samples belong to particle size of 0–1 fi (coarse sand). The mean of gravel sediments in the medial sector of the studied areas is 34%, the mean of sandy sediments is 44% and the mean of glycates sediments is 22%. The study of sediments of the end of the conjugate section (samples of R18, R20) in this area is the most abundant particles of particles belonging to particles larger than 4 fi (cith and clay). The mean of gravel sediments in the distal section of the studied samples is 19%, the mean of sandy sediments is 35% and the mean of glycase sediments is 46% (Fig. 3.2).

3.1.2 *Cumulative Frequency Curve*

This curve is drawn from the cumulative percentage of sediments and the slope of the sediments. The curve with a large slope indicates less standard deviation and better sedimentation. Also, if particles are in size in different groups, the fracture is created in the curve (Pourmorad 2018). The percentage of coarse particles is higher. The parallel curve of the x-axis will be figured, which is a parabolic and is related to the environment, such as the river, which has been very high for a long time, and

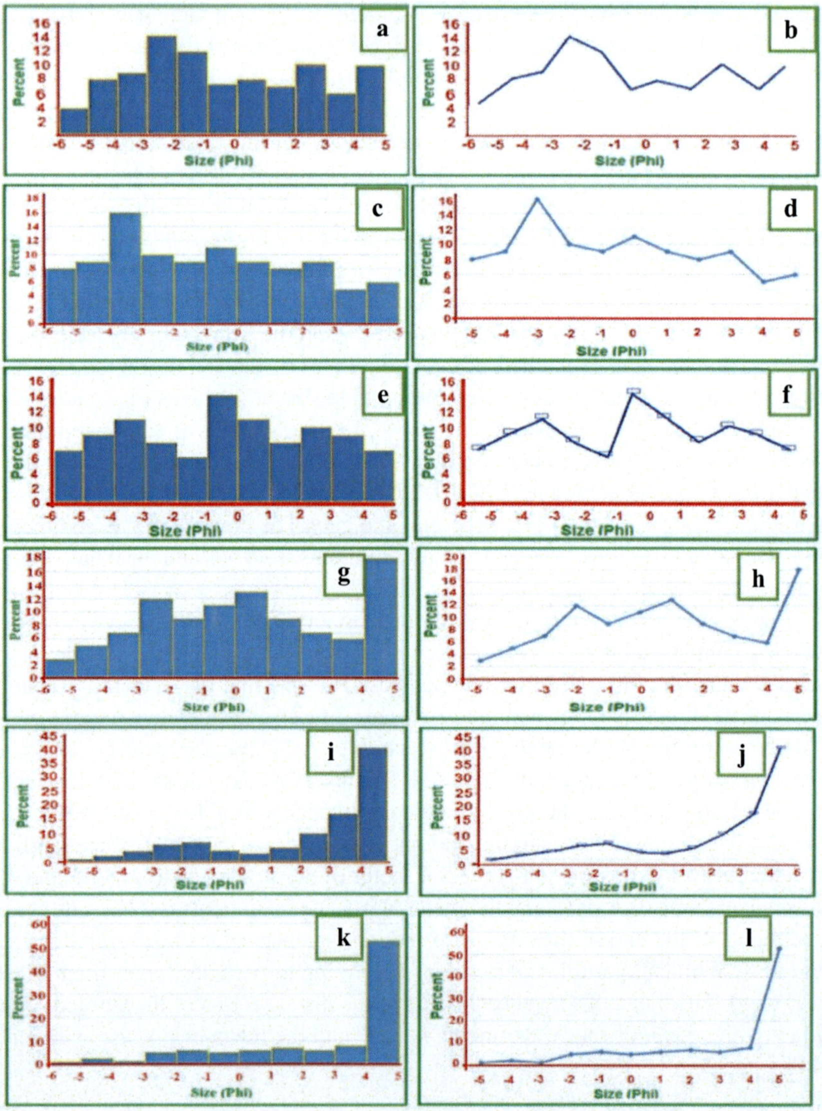

Fig. 3.1 Histogram diagrams and normal distribution of conjugations of Dezful region. (**a** and **b**) Histogram diagrams (**a**) and normal distribution (**b**) from the upper part of the concrete (sample number D5), (**c** and **d**) histogram diagrams (**c**) and normal distribution (**d**) from the upper part of the concrete (sample number D8), (**e** and **f**) histogram diagrams (**e**) and normal distribution (**f**) from the middle part of the concrete (sample D15), (**g** and **h**) histogram diagrams (**g**) and normal distribution (**h**) from the middle part of the concrete (sample number D19), (**i** and **j**) histogram (**i**) and normal distribution (**j**) from the end of the concrete (sample D23), (**k** and **l**) histogram diagrams (**k**) and normal distribution (**l**) from the end of the conjugate section (sample number D26)

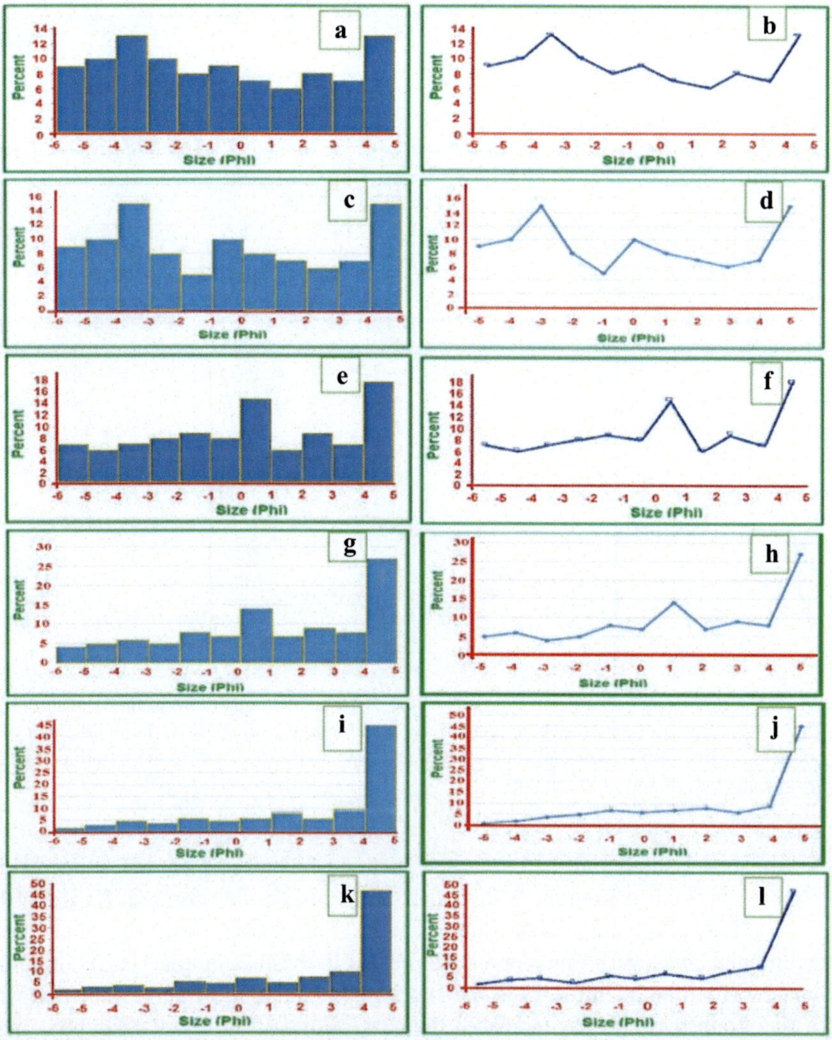

Fig. 3.2 Histogram diagrams and normal distribution of Ramhormoz region. (**a** and **b**) Histogram chart (**a**) and normal distribution (**b**) high section of concrete hand (sample R2), (**c** and **d**) histogram chart (**c**) and normal distribution (**d**) distribution of the hand concrete (sample R6 number) (**e** and **f**) histogram chart (**e**) and normal distribution (**f**) the middle part of the concrete (sample number of R10), (**g** and **h**) histogram diagram (**g**) and normal distribution (**h**) the middle part of the concrete (sample number R13), (**i** and **j**) histogram (**i**) and normal distribution (**j**) distribution of the bottom of the concrete (sample of R18), (**k** and **l**) histogram (**k**) and normal distribution (**l**) of the lower part of the concrete (sample number R20)

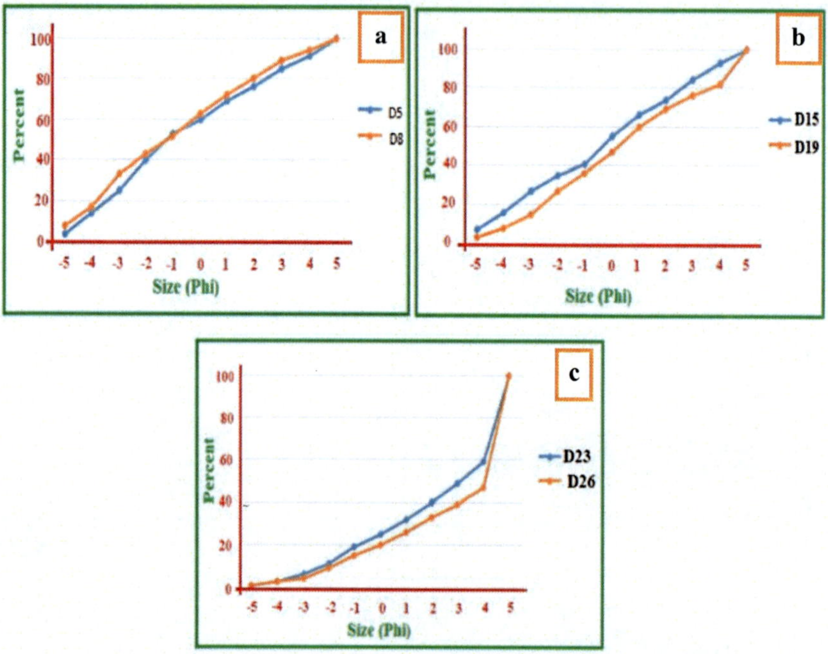

Fig. 3.3 Cumulative curves from the study of the conjugated conjugations of the Dezful region. **a** The upstream area of the area (blue line D5 sample and yellow line D8), **b** the middle part of the region (blue line D15 sample and yellow line sample 19), **c** the lower part of the area (blue line D23 sample and yellow line D26 sample)

sediments are small and the curve in a section of The sediment has a steep slope, namely, energy is continuously high and sediments are well-formed (Karimi 2014).

The study of this curve in the samples of the studied areas shows that both types of ceremonial and logarithmic curves are present in these samples. Also, all diagrams have two to three fractures (Figs. 3.3 and 3.4). In fact, in these curves, the frequency of fine grain particles has increased the curve slope of the studied samples. In the upstream and middle parts of the studied samples, due to weaker formulas compared to the lower parts of these areas, three coarse grain populations, medium grain and fine grain are completely detectable.

3.1.3 *Statistical Parameters*

For detailed study of the studied samples, sorting, skew, middle and other related curves have been drawn, which are examined in the following section.

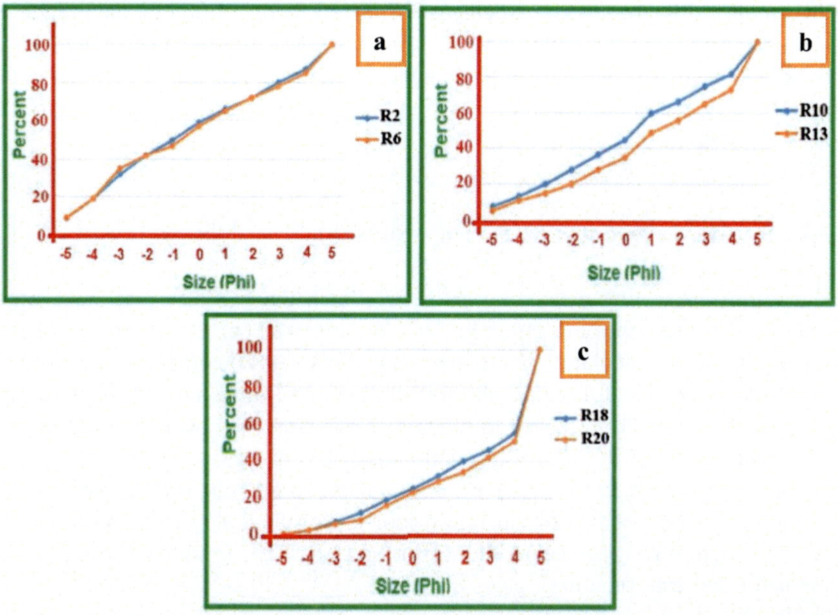

Fig. 3.4 Cumulative curves of Ramhormoz region. **a** The upper part of the conjugal hand (the blue line of the R2 sample and the yellow line of the R6). **b** The middle part of the constian (blue line of the R10 sample and the yellow line of R13), **c** the lower part of the cones (the blue line of the R18 sample and the yellow line of the R20 sample)

3.1.3.1 Median

The Median is the middle part of the curve so that half of the particles are larger than it and the other half are smaller than it (Karimi 2014). Studies conducted in southwestern Iran (Dezful and Ramhormoz regions) show that the median rate of the samples studied in the opposite parts of these alluvial fans is different due to the distance from the source. According to the data obtained from the study of the 3 sections of the region, the mean of the upstream part of this region varies from −3.34 to −1.06, the middle part from −1.13 to 1.69 and the final part from 1.73 to 2.27 (Fig. 3.6a). Also, based on the data obtained from the middle study in two study sections of Ramhormoz region, the average in this region is somewhat similar to the Dezful region and in the upstream part between −3.21 and −1.09, in the middle part −0.81 to 1.07 and in the final part 1.13–1.71 is the median rate of change (Fig. 3.6b).

3.1.3.2 Mean

The Mean can be the average particle size limit of the sediment (Pourmorad et al. 2021). The calculation of average of the alluvial fans of the Dezful region shows that the average of the upstream part changes from −3.34 to −1.06 Phi, the middle part changes from −71 to 1.24 Phi and the final part changes from 1.04 to 2.18 Phi. In the Ramhormoz region, the average changes of the proximal part are from −3.24 to −1.41 Phi, the middle part from −1.25 to 0.67 Phi and the distal part from 0.62 to 2.19 Phi (Fig. 3.7).

3.1.3.3 Measurement of Uniformity (Sorting)

Uniformity or sorting means the extent to which the particles that make up the sediment are the same or different in size (Apolinarska et al. 2020). In these studies, sorting has been calculated by the method of Folk (1980) comprehensive standard deviation. In the Dezful region, the sorting rate in all three studied sections varies between '0.48 and 1.06' (good to medium sorting) in the upstream section, 1.02–1.16 (poor sorting) in the medial section and 0.91–0.82 (medium sorting) in the downstream section. In the Ramhormoz region, the sorting rate in the two cut sections studied varies between '1.78 and 1.14' (poor sorting) in the upstream section, 1.11–0.73 (medium sorting) in the middle section and 0.71–0.52 (good to medium sorting) in the downstream section (Fig. 3.9).

3.1.3.4 Skewness

Tilt is the heterogeneity of the particle distribution curve, which is determined by the sequence of the distribution curve (Eizenhofer 2020; Apolinarska et al. 2020). In these studies, the amount of tilt has been calculated by the method of comprehensive tilt of Folk (1980). The degree of tilt in the Dezful region in the upstream part ranges between '0.62 and 0.16' (tilt toward fine particles), in the middle part ranges between 0.36 and 0.07 (almost symmetrical) and in the downstream part ranges between −0.03 to −1.34 (symmetric to the particles Coarse grain) (Fig. 3.10a). In the Ramhormoz region, in the upstream part, the amount of tilt is between '0.39 and 0.11' (tilting toward fine particles), in the middle part, 0.13 to −0.13 (almost symmetrical) and in the downstream part, −0.14 to −0.24 (tilting toward particles Coarse grain) (Fig. 3.10b).

3.1.3.5 Kurtosis

Kurtosis is the measurement of the sharpness or elongation of a particle distribution curve, which is obtained from the ratio of the sorting of the curve sequence to the sorting of the middle part of the curve (Eizenhofer 2020). In these studies, the Folk (1980) elongation method has been used ($Ku = \frac{\emptyset_{95} - \emptyset_5}{2.44(\emptyset_{75} - \emptyset_{25})}$). The amount of elongation in the Dezful region varies between '1.39 and 1.09' (medium to elongated) in the upstream, 1.46–1.06 (elongated) in the middle and 1.11–0.71 (wide to medium) the downstream. In the Ramhormoz region, the upstream part ranges between 1.34

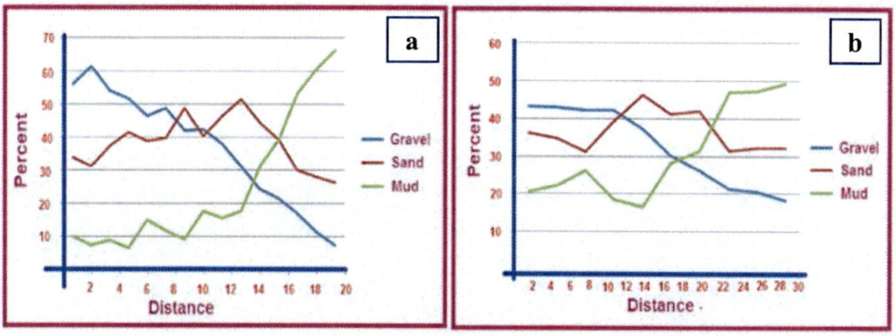

Fig. 3.5 Changes in the percentage of different particles relative to the distance from the apex of the alluvial fan in the studied areas **a** Changes in grain size in section No. 1 of Andimeshk area, **b** changes in grain size in section No. 1 of Ramhormoz area (Ala River), The blue line indicates the percentage of gravel, the red line indicates the percentage of sand and the green line indicates the percentage of silt and clay

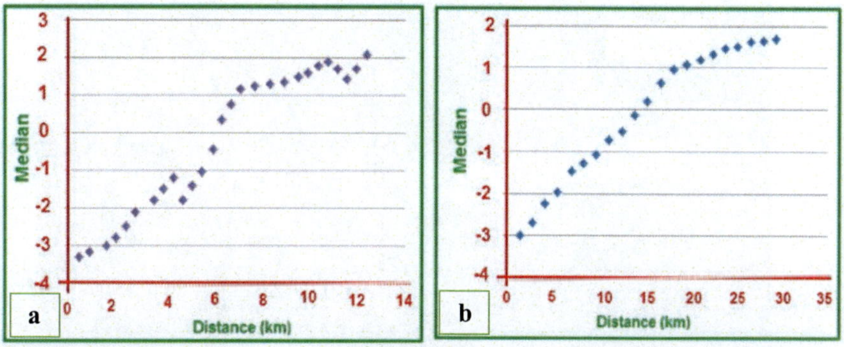

Fig. 3.6 The average amount in the studied samples in relation to the distance from the origin. **a** The average of the studied samples in section 1 of the Dezful regio, **b** the average of the studied samples in section 1 of the Ramhormoz region

and 1.16 (elongated), the middle part between 1.13 and 0.89 (medium to elongated) and the distal part between 0.89 and 0.76 (wide to medium) (Figs. 3.3, 3.4, 3.5, 3.6, 3.7, 3.8, 3.9, 3.10 and 3.11).

3.1.4 Discussion

The particle size of the sediments actually confirms the energy of the transport agent (Chen et al. 2020). If in a mass of sediments, there are many coarse grains (up to gravel), this frequency can indicate the maximum flow velocity during sedimentation (Mitra et al. 2020).

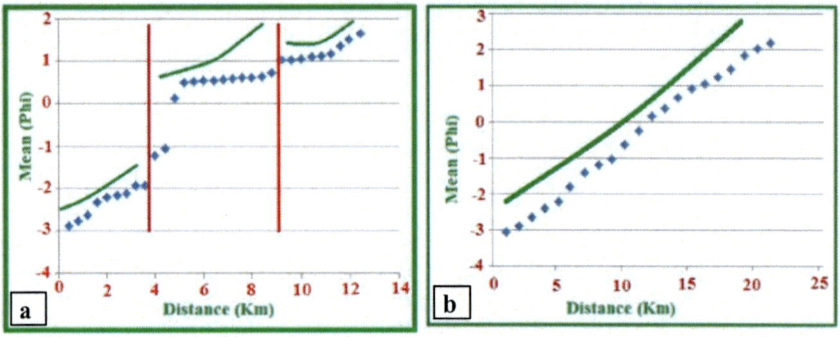

Fig. 3.7 Graph of the average versus distance from the apex of alluvial fans in section No. 1 of Dezful (**a**) and section No. 1 of Ramhormoz (**b**). The green lines in these diagrams show the sedimentary continuity among the sedimentary discontinuities (red vertical lines)

Fig. 3.8 View of the entry of seasonal rivers in the medial section of the cones of Andimeshk, Southwest of Iran

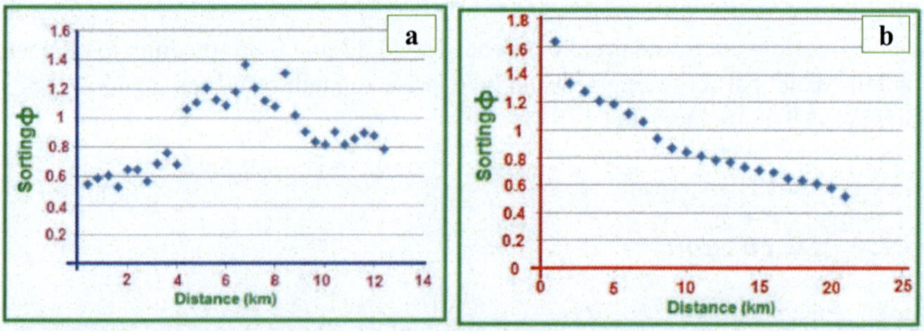

Fig. 3.9 Graph of the sorting rate of the samples against the distance from the apex of the alluvial fans in the studied areas. **a** Cut number one of the Dezful region, **b** cut number one of the Ramhormoz region

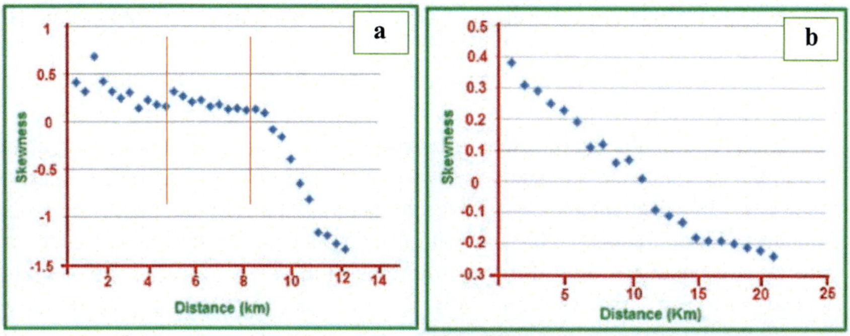

Fig. 3.10 Skewness diagram in front of the distance from the apex of the alluvial fans in the studied areas. **a** Section number one of Dezful (yellow lines indicate sedimentary discontinuities in this area) **b** section number one of Ramhormoz area

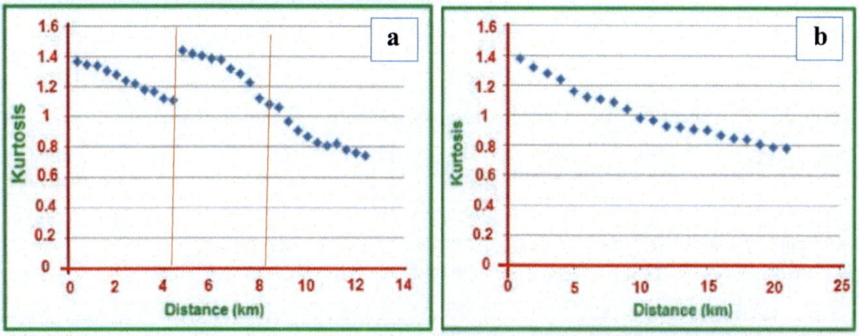

Fig. 3.11 Kurtosis diagram in front of the distance from the apex of alluvial fans in the studied areas. **a** Dezful section (yellow lines indicate sedimentary discontinuities in this area), **b** Ramhormoz section

The particle size distribution of sediments depends on the characteristics of the source rock, weathering processes, wear and their selective sorting during transport (Abu et al. 2020). Hydraulic sorting and abrasion are often the two main factors in the downstream particle size of alluvial fans (Xiao 2016). Geomorphological factors such as reducing the slope of riverbeds along the alluvial fan path is one of the factors controlling hydraulic sorting so that by reducing the slope of the river, the sediment carrying capacity decreases and causes selective settling of large particles and small particles (Stow 2005). Hydraulic sorting depends on the figure, size and density of the particles (Eoff 2014). In addition to hydraulic sorting, particle size also depends on the wear process (Abu 2018). Abrasion is a hybrid process that involves sand blasting, grinding, breaking and sand blasting (Mitra et al. 2020).

The results of the grain size analysis in the samples in the alluvial fans studied in southwestern Iran (in both Dezful and Ramhormoz regions) show that the particle size from upstream to downstream of the alluvial fan shows different changes. So that these sediments are from the type of gravel to sand gravel at the top and in the middle sections with the reduction of particle size to gravel sand to mud sand and these changes in particle size by becoming sandy mud in the section Ends leading to the floodplain (Fig. 3.5 and Tables 3.1, 3.2 and 3.3).

The higher abundance of coarse-grained particles among the finer-grained particles in the Ramhormoz area indicates two important points: first, the flood-carrying sediments in the Ramhormoz area have a higher velocity (and not necessarily higher discharge) Humid and dry climates with the dominance of physical destruction. In contrast, the median study of these sediments in the Dezful region shows that the currents carrying them were relatively slow but more continuous; in such a way that the floods carrying them were not in the form of showers, but occurred due to continuous rains in a humid climate.

The study of cumulative frequency curves of the samples related to the studied areas shows that, in the upstream and middle parts of the studied samples due to weaker sorting than the samples in the downstream parts of these areas, three of the statistical population of coarse-grained, medium-grained and fine-grained are quite distinguishable. The results of the study of cumulative frequency curves of

Table 3.2 Results of the statistical parameters of southwestern Iran (Dezful era)

D8	D7	D6	D5	D4	D3	D2	D1	Sample number
32° 26′ 57″–48° 27′ 36″ E	32° 27′ 2″–48° 27′ 54″	32° 27′ 30″–48° 28′ 27″	32° 27′ 30″–48° 28′ 27″	32° 27′ 22″–48° 29′ 4″	32° 27′ 24″–48° 29′ 30″	32° 27′ 48.38″–48° 29′ 54.49″	32° 28′ 34″–48° 30′ 7″	Location
−2.3	−2.1	−2.5	−2.8	−3	−2.9	−3.2	−3.1	Median
−2.13	−1.95	−2.18	−2.21	−2.34	−2.64	−2.78	−2.9	Mean
0.14	0.31	0.25	0.32	0.42	0.68	0.32	0.41	Skewness
0.69	0.57	0.65	0.65	0.53	0.61	0.59	0.55	Sorting
1.18	1.22	1.24	1.28	1.31	1.34	1.35	1.37	Kurtosis

Table 3.3 Results of the statistical parameters of southwestern Iran (Ramhormoz era)

R8	R7	R6	R5	R4	R3	R2	R1	Sample number
31° 14′ 12″–49° 35′ 47″	31° 14′ 25″–49° 36′ 20″	31° 14′ 34″–49° 36′ 42″	31° 14′ 44″–49° 36′ 56″	31° 14′ 56″–49° 37′ 18″	31° 15′ 4″–49° 37′ 34″	31°15′ 9″–49° 38′ 33″	31° 15′ 10″–49° 39′ 32″	Location
−0.67	−1.19	−1.29	−1.36	−2.11	−2.24	−2.71	−3.01	Median
−1.18	−1.41	−2.09	−2.27	−2.32	−2.65	−2.91	−3.07	Mean
0.12	0.11	0.19	0.23	0.25	0.29	0.31	0.38	Skewness
0.94	1.06	1.12	1.19	1.21	1.28	1.34	1.64	Sorting
1.17	1.18	1.17	1.2	1.22	1.27	1.32	1.38	Kurtosis

the studied areas indicate the presence of high slope fractures in the end sections of the curves obtained from the samples of the middle and end sections of the alluvial fan cones (Figs. 3.3b, c and 3.4b, c). The presence of fractures in these part of the diagrams indicates the abundance of fine particles in this part of the alluvial fans due to moving away from the top of the studied alluvial fans, thus reducing the ambient energy to carry larger grain particles in these parts (Figs. 3.3 and 3.4).

Also, the study of these sediments in the middle part of the Ramhormoz region shows that two climatic transition stages have occurred from the wet to the relatively dry period. Thus, the median of sediments in this region, especially in their upstream part, has the lowest numerical value, and this indicates that the particles have occurred by floods with strong energy but less persistence. This result shows that these sediments are more as a result. Flood flows have been deposited (Fig. 3.6).

One of the appropriate methods to investigate the amount of fineness in environments with sedimentary discontinuities is to divide the study area into separate sedimentary linkages based on the identified discontinuities (Mitra et al. 2020). In relation to sedimentary discontinuities, several factors such as sedimentary load, lithological changes, type of rivers in the region, climatic conditions, geomorphological phenomena and tectonic activity have an important role.

In sediments of the Dezful region, two sedimentary discontinuities and 3 sedimentary continuities have been detected. The first discontinuity is near the apex of the alluvial fan and is related to the changes in the slope of the area from the steep section (upstream of the area) to the lower slope section (middle section of the alluvial fan). The second discontinuity, which is in the middle parts of the alluvial fan, belongs to the entry of seasonal rivers into this part of the alluvial fan (Fig. 3.7a). The inflow of these rivers causes the coarse-grained parts of the Bakhtiari Formation to be transported to this section, and as a result, the average particles in this section change (Fig. 3.11). In the Ramhormoz region, due to the large area and gentle slope changes in the region and the lack of significant lithological changes in the region, no significant sedimentary discontinuity is observed and the particle shredding process is gradual and has the same sedimentary continuity (Fig. 3.7b).

By comparing the sorting rate in Ramhormoz and Dezful regions, with the exception of the proximal part, which is better sorted in the Dezful region due to several sedimentary cycles and the nature of its Bakhtiari source rock, we conclude that, in general, the sorting process of particles in Ramhormoz region is more regular than Dezful region from the top to the end of the alluvial fans.

According to the presented data, the rate of sorting changes in Dezful region is observed with more fluctuations than Ramhormoz region (Fig. 3.9). In the Ramhormoz area, due to changes in the grain size from the apex to the downstream and the absence of sedimentary discontinuity, regular changes from poor sorting (at the apex of the alluvial fan) to relatively good sorting (downstream) are observed, but in the Dezful region, due to the presence of sedimentary discontinuities that were discussed in the section on the average particle size, we have different changes in particle sorting. In fact, sharp fluctuations in particle sorting can be attributed to sedimentary discontinuities along the canal, such that after a sedimentary cohesion that occurs in a finite process by one of the factors causing the discontinuity (including the entry of coarse-grained particles through a stream). In the Dezful region, particles with an average diameter more than the sediment particles enter the main channel of the river at the point of discontinuity and decrease by mixing particles with different diameters.

In Dezful region in Ras section, coarse gravel particles which are observed as dominant particles have relatively good initial sorting due to the nature of Bakhtiari Formation. These sediments are expected to gradually increase in size after transporting them to the downstream of the alluvial fan, but along this path, two important sedimentary discontinuities change the particle size (Fig. 3.9).

Studies show that the data obtained from the Skewness parameter are consistent with the data obtained from the study of mean diagrams versus distance and indicate the existence of two discontinuities in the Dezful study area (Fig. 3.10a). In general, the amount of tilt shows a positive trend by approaching the middle and downstream parts and with the abundance of fine particles in these two areas. The degree of Skewness of the Dezful region in the upstream part is toward fine-grained particles, the middle part is symmetrical and the downstream part is symmetrical up to coarse-grained particles (Fig. 3.10a). In the Ramhormoz region, in the upstream part, the amount of inclination is toward fine-grained particles, the middle part is almost symmetrical and the downstream part is symmetrical up to coarse-grained particles (Fig. 3.10b). The presence of positive skewness in these alluvial fans along with negative sorting indicates the presence of relatively calm to relatively severe river currents in different periods, so that these currents have been able to settle these particles of different sizes. In addition, the positive distortion is due to the presence of large amounts of suspended solids, such as silt and clay, in the river, some of these particles remain in the sediments after deposition. Rivers that carry large amounts of coarse-grained particles may be negatively skewed.

In the Dezful region, the amount of sediment Kurtosis in the initial parts of the alluvial fan, where the frequency of coarse-grained particles is much higher than fine-grained particles, is moderate to elongate. In the middle parts, the amount of Kurtosis is seen as elongated due to the discontinuities mentioned earlier, which gradually move to the lower part due to the abundance of fine particles, moving away from the middle part to the downstream parts. It is observed more widely in Fig. 3.11a. In the Ramhormoz area, due to the lack of significant discontinuities in the study area and gradual changes in particle size from upstream to downstream of the studied alluvial fan, elongation at first in the upstream sections due to the greater abundance of coarse-grained particles than grain particles. Fine, elongated and gradually approaching the middle and end parts of the alluvial fan due to the greater abundance of fine-grained particles than larger grain particles, is seen as medium to wide (Fig. 3.11b).

From the overall data, it can be inferred that in both Dezful and Ramhormoz regions, the particle size from upstream to downstream shows large changes in terms of particle size, so that at the top of both regions large particles and downstream particles. There are fine silt and clay grains. But what distinguishes the sediments of Dezful region from the sediments of Ramhormoz region is the existence of two sedimentary discontinuities in the alluvial fans of this region.

In fact, these discontinuities cause changes in grain size, resulting in mean, median, sorting, and somewhat skewed graphs from the study of these sediments, while in the Ramhormoz area, where sediments are mainly transported by the main river. And secondary factors such as the entry of sub-canals have less effect on their size changes. China is not in a state and witnessed a sedimentary order in terms of grain size changes, average, middle, sorting and Skewness from the top of this alluvial fan downstream that we have witnessed.

3.2 Study of Sedimentary Facies

Sedimentary facies differs from other parts of a stratigraphic unit in terms of lithological and paleontological features (Ghorbani 2019). If the stratigraphic unit is divided based on rock data and minerals, it is called lithofacies, and if the stratigraphic unit is divided based on fossils in the rock, it is called biofacies (Tuker 2001). According to Selley (1996), the important point in interpreting sedimentary environments is the correct and accurate understanding of the concept of sedimentary facies. A sedimentary facies is a collection of sedimentary rocks that has a geometric form (geometry), lithology (lithology), sedimentary structures, long flow pattern and certain fossils with the same characteristics of the rock. Others are distinguishable (Selley 1996). Studies of sediments and sedimentary rocks can be followed at different scales, especially in the analysis of sedimentary facies, which can be divided into two general groups: macroscopic and microscopic (Ghorbani 2019).

Depending on the scale of study, lithology, the type of available data and other characteristics, sedimentary facies types are common in sedimentary geology today, some of which are: rock facies (lithofacies), microscopic facies, Petrographic facies, log facies (electrofacies), biological facies (biofacies), ultrafacies (microscopically determined facies (SEM), seismic facies and tectonic facies (tectofacies), etc. (Tuker 2001). In the following section, the rock facies that are the main purpose of this chapter are explained.

3.2.1 Stone Facies or Lithofacies

This facies is defined during macroscopic studies mainly in the desert (outcrops and hand samples) and drilling cores. In fact, rock facies are defined based on texture and sediment structure in sediments and detrital rocks (Miall 2014). In this section, Miall codes (Miall 2014) are used to study the rock facies identified in the two study areas, i.e. Ramhormoz and Dezful, where a total of 11 rock facies have been identified and examined in the following section. The interpretation of these facies is discussed in Table 3.4.

Table 3.4 Name and description of facies identified in Ramhormoz and Dezful regions

Code	Lithofacies	Interpertation
Gmm	Matrix supported massive congelomerate	Gravel deposits with mass structure resulting from plastic retail streams (viscose, high energy)
Gmg	Matrix supported graded gravele	Deposits with positive gradual grain structure resulting from high sediment load flows in the area near the source
Gcm	Clast supported massive gravele	Gravel deposits with mass sediment structure without special arrangement, resulting from particle flows with false plasticity and high sediment load
Gci	Clast supported gravel	Deposits from low viscosity particulate streams
Gp	Planar cross bedded gravele	Gravel deposits of support grains due to migration of longitudinal and lingual dams in deep sections of the canal
Gh	Horizontally bedded gravele	Grovele parts, rich in parts and with poor horizontal stratification due to energetic tensile currents with shallow depth and high concentration of sediment
Sm	Massive sandy	Sand deposits with mass structure resulting from gravitational currents
Sh	Horizontally bedded sand	Flat sand sheets in the upstream and downstream regime above the ridges or inside the canal
Sp	Planar cross bedded sand	Deposits from the migration of two-dimensional soils with a wave ridge in the downstream regime
FM	Massive mud and silt	Deposits from flow energy reduction in isolated canals and flood zones
Fi	Laminted mud and silt	Mass silty sediments resulting from the deposition of suspended particles due to reduced flow velocity in the area outside the channel and above the sand ridges

Since this book is written for Ph.D. students, introductory discussions such as the introduction of sedimentary facies introduced by Miall (2014) have been avoided and only facies observed in the region in southwestern Iran have been introduced.

In fact, the main purpose of this book is practical and advanced teaching, therefore, the presentation of basic materials have been avoided.

The rock facies identified in the Dezful and Ramhormoz alluvial fans are classified into three categories of coarse, medium and fine grains (based on Miall classification, 2014) based on grain size, which are described in the following section.

3.2.1.1 Coarse-Grained Rock Facies (Gravel)

The most important coarse-grained facies identified in the two study areas including Gmg, Gmm, Gp, Gcm, Gh, Gci are the following that examines the frequency of each facet in the studied alluvial fans.

Mass Matrix-Rich Gravel Facies (Gmm)

These facies form the most abundant coarse-grained facies in the studied alluvial fans.

These facies are in the form of thick layers that have a massive figure, and due to their proximity to the source, the grains are relatively angular and their gravels in very large to small sizes (Bauder to Pebble) are included. To be. This facies is mainly composed of a matrix of sand and mud.

In the Dezful region, these facies mainly consist of semi-rounded grains in sizes between 5 and 30 cm and sometimes with embryos.

In this area, these facies are usually along with rock facies of mud and silt mass (Fm) with sudden or erosive contacts. In the Ramhormoz region, these facies have semi-angled grains with sizes between 3 and 22 cm. In the mentioned area, these facies are generally together with the gravel facies of the grain supporting masses (Gcm) and the facies of mud and silt masses (Fm) with sudden lower and upper contacts. These facies in this area are massive and contain a large amount of mud matrix that in Dezful area the amount of mud matrix is less detectable than Dezful area (Fig. 3.12). These facies are observed in terms of frequency in both regions of almost similar expansion and frequency in the upstream parts of these alluvial fans, but as it moves away from the upstream in Ramhormoz alluvial fans, their width and expansion decreases. While these facies are visible up to the downstream parts and even the middle parts of the Dezful alluvial fan due to the main source of sediment transport (retail streams).

These facies in the study areas are between 60 cm to more than 10 m thick (in the canal walls) (Fig. 3.12).

Interpretation of Gmm facies

Massive properties, abundant matrix and uncertain texture of the strata indicate rapid sedimentation periods of the downstream region, where the sediment concentration is high and sedimentation occurs abruptly (Boggs 2018).

The lack of sedimentary structures and the presence of poorly sorted grains also indicate the deposition of mass flows. In addition, the absence of gradual stratification and the presence of a disordered structure in these facies in the study areas indicate the short transport and proximity to the source of sediments of these facies (Miall 2014). From the total information obtained, it can be concluded that the sediments of these facies, which are generally observed in the upstream parts of the target areas, have been created as a result of a period of rapid deposition of small streams.

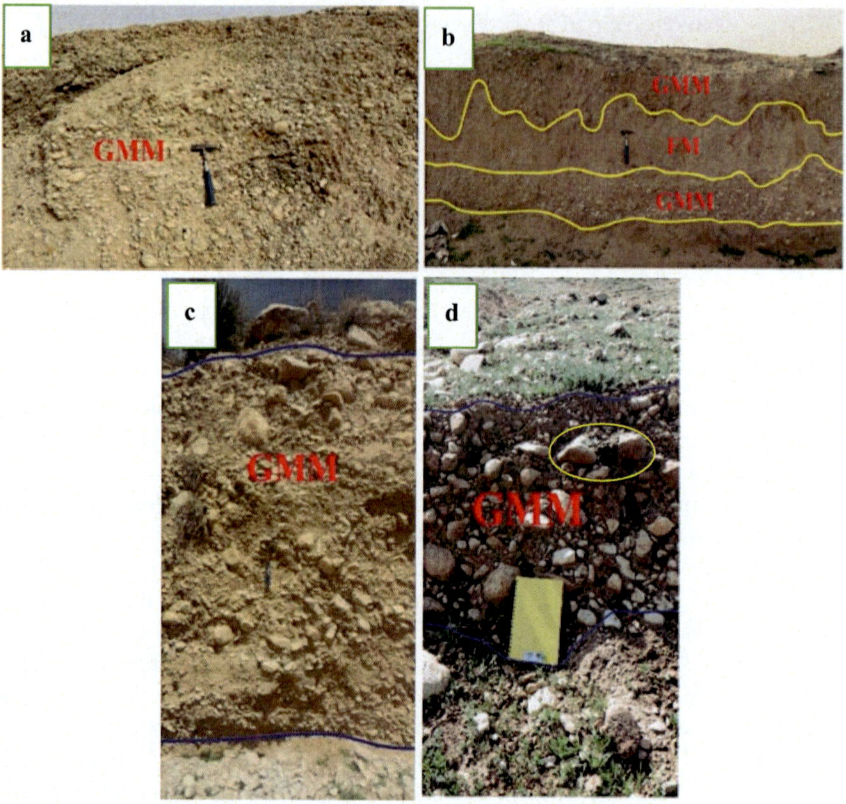

Fig. 3.12 Gmm facies in the studied areas. **a** Gmm facies with relatively poor sorting of parts in the northwest of Dezful, **b** erosion and slope boundaries between Gmm and Fm facies in the north of Shushtar, **c** coarse gravel facies in the northeastern part of Ramhormoz and **d** close view from Gmm facies with coarse-grained gravel parts with embrification (yellow circle) in the northeast of Andimeshk area

Matrix Rich Gravel Facies with Gradual Classification (Gmg)

These facies are generally observed in the Dezful region and are less visible in the Ramhormoz region (Fig. 3.22). The most important feature of these facies in the studied areas is the absence of a rich framework of the piece, which is connected by a matrix of sand, silt and clay. These facies have a normal gradual classification with parts in different sizes from 5 to 60 cm, which includes surface deposits close to the source. The pieces are generally semi-rounded to angled and with relatively poor sorting. In the Ramhormoz region, where particulate currents are the main factor of transportation, these gravel particles have relatively weak sorting and are semi-angular to semi-rounded. The lower and upper contacts of these facies are usually gradual and sometimes erosive (Fig. 3.13).

Fig. 3.13 Facies gravel matrix-rich facies with gradual classification (Gmg) in the alluvial fans of the study area (northwest of Dezful) which shows the upward fineness structure

Interpretation of Gmg facies

The properties expressed in relation to these facies indicate the formation of these facies by high strength retail flows and false plasticity (Blair and McPherson 1999). Also, the existence of normal gradual classification is the result of decreasing flow energy over time. This classification is mainly visible by moving away from the upstream parts and more in the middle parts of the alluvial fan, which is caused by transport over time. And are mainly visible in the small alluvial fans of the Dezful region.

Segment-Rich Gravel Facies with Inverse Gradual Classification (Gci)

In the study areas, the main framework of these facies consists of granular parts in the size of granules to Pebbles. The grains are poorly sorted and semi-rounded.

No special stratification is seen in the mentioned facies and it has a matrix of medium to fine-grained and silty sand deposits. This facade is 2–2.5 m thick in the west of the Dezful region and about 15 m thick in the east of the Ramhormoz region.

Fig. 3.14 Rich section-rich gravel facies with gradual reverse classification with gradual lower contact in west Dezful

In the desired areas, these facies are generally light gray in color and has a gradual lower contact and have embryos (Fig. 3.14).

Interpretation of Gci facies

These facies have been left as deposits of particle streams with low mobility and high adhesion (Miall 2014). In fact, the existence of a gradual inverse classification in these facies indicates the pressure distribution in the flow, which reduces the viscosity and increases the shear stress of the flow, and thus the migration of large parts to higher parts.

Gravel Gravel Facies of Mass Support (Gcm)

One of the main features of these facies is the presence of large gravelly grains (5–24 cm) in a mass without special arrangement next to each other. In these facies, the amount of matrix is very small and includes detrital grains as much as sand. In the Dezful region, these facies are present in the upper part of the alluvial fan and

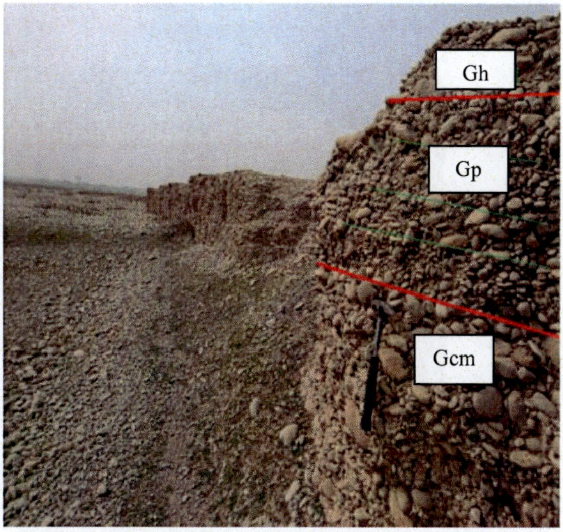

Fig. 3.15 Sudden upper contact of Gcm facies with Gp facies in the north of Andimeshk city (primary source)

the sorting of grains is moderate and semi-circular, which as mentioned before, are re-transported from Bakhtiari Formation (Fig. 3.15).

These facies can be seen in the Ramhormoz alluvial fans, especially in the east and northeast of this city, much more often than the Dezful alluvial fans. These facies have been observed in the eastern part of the Ramhormoz city with poor sorting and relatively weak rounding, which is more visible below the heights of the Gachsaran Formation. In the Dezful region, the maximum particle size is about 24 cm, but in the Ramhormoz region, it is less than 15 cm in the upstream part.

The lower and upper contacts of these facies are usually sudden with other facies and especially with Gh facies (Fig. 3.15). The thickness of these facies varies between 50 cm to more than 20 m in the studied areas (especially in the canal walls). This thickness is the highest in the upper parts of the Ramhormoz River and the least in the proximal parts of the Andimeshk region.

Interpretation of Gcm facies

The presence of grains with weak sorting properties indicates streams with weak turbulence and high density that are in the mass state and are mainly formed in areas close to the source (Arzani 2004). In fact, this facies shows the high energy of the environment during deposition, which prevents the deposition of fine particles. This property is abundantly visible in the eastern and northeastern parts of Ramhormoz city. These facies are probably formed by the transport of bed load in the form of residual channel deposits and flows with high velocity and energy (Boggs 2018). These features can be seen in Ala Ramhormoz River which is a steep river with high flow intensity and also in the northwestern sediments of Andimeshk city which are mainly formed by seasonal river flows.

Fig. 3.16 GP facies in the studied areas. **a** GP facies north of Andimeshk city, **b** sloping border between two facies GP and Gmm in the northeast of Ramhormoz city (primary source)

Gravel Facies with Flat Diagonal Stratification (Gp)

In the study area, these facies are supported by grains with a low angled flat oblique stratification and are brown to dark gray (Fig. 3.16). The thickness of each category of oblique stratifications varies from about 30 cm in the region in Dezful region to more than 10 m in the Ramhormoz region. The grain size of these facies in the Ramhormoz region has changed from 3 to 10 cm with a sand matrix, but in the Dezful region, the size of these grains is between 5 and 12 cm with a sand and silt matrix. In the Ramhormoz region, the parts are mostly semi-angled to semi-rounded and have poor sorting, while in the Dezful region, rounding and sorting of parts are relatively good (Fig. 3.16a). These facies are associated with Gp facies in the studied areas, especially in the Andimeshk region (Fig. 3.16b). The base of these facies are erosive and sometimes gradual.

Interpretation of Gp facies

The rich texture and fabric of these facies in the studied areas indicate the deposition of these facies by energetic tidal currents with shallow depth and high concentration of sediment (especially in the Andimeshk region by seasonal rivers) that transport gravel particles as Bed loading and sedimentation have occurred under flow reduction conditions in the target areas (Boggs 2018). In fact, these facies are formed diagonally due to the movement of two-dimensional megaripples or the deposition of large parts in the riverbed. Also, the formation of flat diagonal layers in these facies, especially in the northwestern part of Dezful city, indicates a slower transport of gravel particles, which causes the vertical elevation of these facies.

Fig. 3.17 Gh facies in the studied areas. **a** Gradual boundary between Gh facies and Gmm facies in the east of Ramhormoz city, **b** Gh facade with slope with facies, Gmm Fm, in the east of Ramhormoz city, **c** sloping boundary between Gh facies with Fm and Gmm facies With grain embryos in the northeast of Andimeshk city

Horizontal Gravel Facies (Gh)

Gh facies consists of semi-circular and relatively assorted parts, rich in fragments with poor horizontal stratification. These facies have a low frequency in alluvial fans of the Dezful region, but it has a much higher frequency in alluvial fans of the Ramhormoz region due to the high intensity of river transport, especially near the river. The grain size in the Ramhormoz area is usually about 5–9 cm with relatively weak sorting and semi-rounded (Fig. 3.17a, b). However, in the Andimeshk region, the size of these grains is between 4 and 11 cm with relatively weak sorting and are semi-rounded to angular with impregnation visible (Fig. 3.17c). The grains of

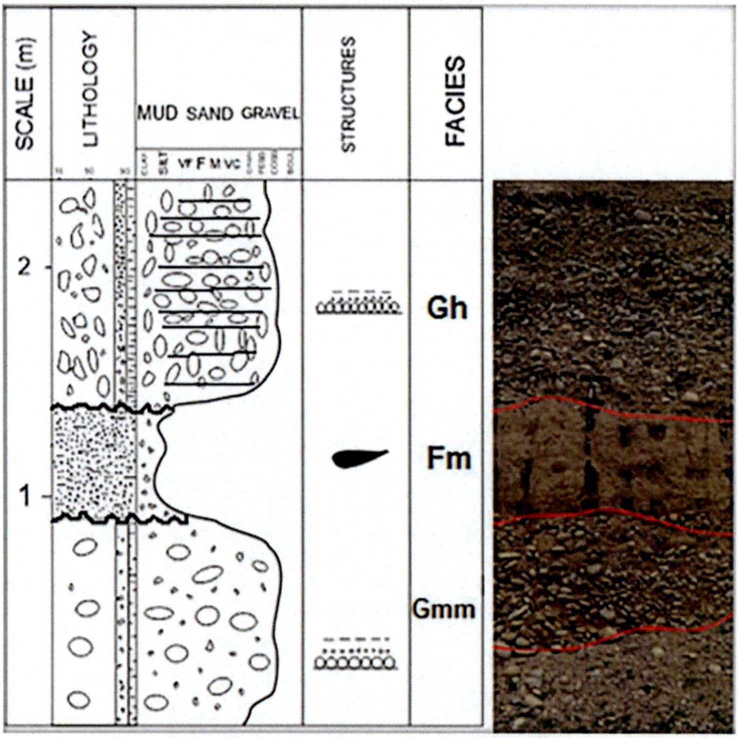

Fig. 3.18 Erosion boundary of Gh facies with Fm facies with relevant stratigraphic column (north of Andimeshk, southwest of Iran)

these facies are generally associated with Gmm and Fm facies in the study areas. The thickness of these facies varies between 50 cm and 8 m in the study areas and the matrix is between pieces of sandy and silty sediments. The highest frequency of these facies is observed in the eastern part of the Ramhormoz region and in the canal walls. The lower and upper contacts of these facies are erosive and sometimes gradual with the mentioned facies (Fig. 3.18).

Gh facies interpretation

The rich texture and fabric of these sedimentary facies show that the deposition of these facies by energetic tidal currents with shallow depth and high concentration of sediment in the study areas where the transport of gravel particles is done as bed load and sediment. The deposition occurred under flow reduction conditions (sometimes in the form of channel-filling sediments). Since sediment supply is high, sediment uplift has been done laterally and downstream of faster flow, which has caused horizontal stratification in these sedimentary facies (Miall 2014).

Fig. 3.19 Lateral expansion of Sm facies. **a** Expansion of these facies in the north of Andimeshk with medium to weakly sorted grains with erosion boundary, **b** alternation of sandy sediments with gravel layers in the middle of Ramhormoz alluvial fan with erosion boundary. Massive sand (Sm), support grain conglomerate (Gcm) and construction of excavation and filling at the base (Sf)

3.2.1.2 Medium Grain Rock Facies

The most important medium grain facies in both Dezful and Ramhormoz regions include Sm, Sh and Sp, which are described below.

Sand Facies with Mass Classification (Sm)

These facies, which include fine-grained to coarse-grained sands, are one of the most abundant medium grained facies in Ramhormoz and Dezful regions. The grains of this facies are moderately to weakly and semi-rounded and are mostly the result of erosion of conglomerate and sandstone units of the catchment (Fig. 3.19a). These facies are seen in the mentioned areas without internal structure, sometimes with horizontal layering and light gray to dark color, and sometimes with protruding and filling structures at the base and with a clay matrix (Fig. 3.19b).

In the study areas, these facies are between 30 cm and 5 m thick. The highest thickness of these facies is in the Andimeshk region and its lowest thickness is in the southeast of the Ramhormoz region. These facies are generally observed along with the Gcm facies in the studied areas and the lower and upper contact of these facies with the mentioned facies is generally observed as erosion (Fig. 3.19).

Interpretation of Sm facies

The presence of these facies with mass characteristics indicates a rapid rate of sedimentation in the study areas (Miall 2014). These facies may have been deposited due to the high rate of sedimentation and under the influence of gravitational currents in the study areas. Defective layering or its absence can be attributed to the abundance of sediment supply as well as streams with high sediment load. The erosion of the lower part of these facies indicates the fluctuations of flow in these areas, which is consistent with the fluctuations of river flows, especially in the Ramhormoz region.

Fig. 3.20 Sandstone facies with horizontal layers with sudden boundary with Fm facies in the north of Andimeshk, southwest of Iran

Sandstone Facies with Horizontal Layers (Sh)

These facies in the study areas is composed of medium-grained to coarse-grained sand sediments with medium sorting (in Ramhormoz region) and relatively good (in Dezful region). The thickness of these facies in the studied areas varies from 30 cm to more than 3 m with its highest thickness in the Andimeshk region and its lowest thickness in the southeast of the Ramhormoz region along with the mud matrix. The lower and upper boundaries of these facies are abrupt and these facies in the Andimeshk area is usually observed laterally with sandy facies showing mass classification (Sm) (Fig. 3.20).

Interpretation of Sh facies

The main feature of this facies is the presence of parallel and horizontal layers and laminations that the layers are at the level of lamination (Hosseini Barzi and Bayet Goll 2010). These facies can be formed in sediments of different sizes and at different velocities of water flow, so that, in high currents, laminations are mostly formed in sediments of sand and silt, and in general, these facies represent one-way, low-energy traction currents (Miall 2014). In the studied areas, high-energy river conditions, especially in the case of Ala Ramhormoz River, can be considered as one of the main factors in the formation of these facies.

Fig. 3.21 Sp facies, which are located by sudden contact below the Fm facies (along with plant roots) and are separated by black lines (southeast of Ramhormoz)

Sandstone Facies with Flat Diagonal Layers (Sp)

In the study area, these facies consist of medium to large sand particles with flat diagonal layering showing low angle and relatively good sorting and semi-rounded. The thickness of these facies in the studied areas varies between 30 cm and 1 m. The lower and upper contacts of these facies in the studied areas are usually sudden or erosive, and in these areas, these are generally observed along with the Fm facies. These facies in the Ramhormoz area are mainly observed as light gray to reddish color (Fig. 3.21).

Interpretation of Sp

These facies can be further formed by the movement of two-dimensional ripples with a straight ridge (Boggs 2018). In fact, the existence of such facies indicates a flow with a relatively quiet and intermittent period of activity in the area.

3.2.1.3 Fine-Grained Rock Facies

Fine-grained facies in the study area are mainly composed of mud and silty facies and sandstone facies and laminate silt. These facies are mostly found in the end sections of the alluvial fans, which are explained in the following section.

Facial Mud and Silt Mass (FM)

In the study area, these facies consist of mud and silt with a mass state without any layering and sedimentary structure and are usually observed in the study area as light gray to dark gray (Fig. 3.22a, b). These facies, which are abundantly visible in most of the studied areas, are generally visible along with the roots of plants (Fig. 3.32c). In the study areas, these facies have a thickness between 30 cm to more than 10 m.

Fig. 3.22 Expansion of Fm facies in the study area. **a** Study area (east of Andimeshk), **b** Fm facies in Ramhormoz alluvial fan (south of Ramhormoz), **c** sudden and erosive border of Fm facies with Gcm and Gmm facies (east of Dezful city, southwest of Iran)

This facies are generally observed with Gmm and Gcm facies. The lower and upper contacts of these facies in these areas are usually observed suddenly or erosively (Fig. 3.22c).

Sand Facies-Mud and Silt Laminate (Fi)

In the study area, these facies include a collection of fine-grained sediments of rock mud that the size of the grains is in the clay but in some parts of the studied sequences are in the sand or silt. The thickness of these facies in the studied areas varies between half a meter to more than 8 m (Fig. 3.23). These facies in the study area sometimes have traces of plant roots and the lower and upper contacts of these facies are usually erosive and sometimes gradual. In the studied areas, the mentioned facies has more frequency in the southeast of Ramhormoz region and the lowest frequency in the Dezful region.

Fig. 3.23 Fi facies in the studied areas. **a** Fi facies with plant roots in the southeast of Ramhormoz, **b** Fi facies in southeast Andimeshk, southwest of Iran

Interpretation of fine-grained facies of the region

These facies represent deposits resulting from the subsidence of strong flood currents that the flow strength has decreased sufficiently for the deposition of destructive sediments the size of the herd (Barkashani et al. 2014).

In the study area, in addition to the fact that in some parts the roots of plants have caused the layering in these facies (mostly in the Dezful region), strong flood currents have played a role in the formation of these facies, especially in the southwest of Ramhormoz.

In fact, the Fm facies, which consist of silty and clayey sediments in a massive state without any specific lamination, indicates strong flood currents, which indicate that the flow capacity for depositing destructive sediments the size of mud and silt from the suspended stream has decreased. These facies, which are more abundant in the southern regions of Andimeshk alluvial fans, have been formed due to the reduction of flood velocities and as a result of the deposition of these sediments in this region.

In addition, these facies in the Ramhormoz area can be formed as a result of rapid deposition of suspended particles in water in an energetic environment such as the river environment.In addition, the presence of Fi facies in this area and the presence of flat laminations in these facies can be analyzed such that the sediments of these facies in the study areas are mostly obtained at low energy of water flow and as a result of suspension currents. In fact, the sediments of these facies are often suspended, which may be left behind by weak tensile currents.

3.2.2 Structural Elements

The set of rock facies can represent the separation of structural elements (Miall 2014). Structural elements include sedimentation and erosion forms that are classified based on the sedimentary environment (Fayazi Borujeni et al. 2017). In order to identify

Table 3.5 Summary of properties and facies of structural elements studied in Dezful and Ramhormoz regions based on Miall classification (2014)

Structural elements	The observed sedimentary facies assemblages	Interpretation
GB	Sm, Gmm	It is the result of the migration of gravel barriers or is in the form of residual sediment in the channel floor; it is usually lenticular in figure
CH	Gp, Gcm, Gh, Gmm, Sh	It represents channel filler sediments; the lower boundaries are erosional; generally, it is lenticular and widespread
SG	Gmm, Gmg, Gcm, Gci	Debris flows are adjacent to the provenance area; they have thick sheet-like structures, erosional boundaries, abundant coarse gravels, poor sorting and lack of stratification
SB	Sh	sheet-like, widespread; usually, they are in the form of channel filler sediment, locating in downstream
FF	Fl, Fm	It contains off-channel fine sediment, in massive and sheet-like forms

the structural elements in the study areas, features such as the type of lower and upper contact surface of the layers, the geometric form of sediments, lithology, the direction of flow and the set of their facies have been considered.

The five identified structural elements include channel-filling sediments (CH), dams and gravelly layers (GB), deposits of gravitational sediment flow (SG), sandy layers (SB) and Fine-grained sediments are floodplain (FF). These elements are defined based on their geometric figure and enclosing surfaces (Miall 2014) in order to interpret the sedimentation environment. The structural elements identified in Ramhormoz and Dezful regions are presented in Table 3.5.

3.2.2.1 Gravel Bars and Bedforms Structural Element (GB)

This structural element includes layered figures and lenticular gravel dams that are common in rivers with gravel bed load (Miall 2014). This structural element in the study areas is composed of Gmm and Sm sedimentary facies and is widely formed at the bottom of the canal due to the migration of longitudinal dams upstream of the river. The lower contact is usually either sloping or eroded (Fig. 3.24). The thickness of these structural elements varies between 1 and more than 10 m. The lower and upper contacts of these structural elements are gradual and sometimes erosive.

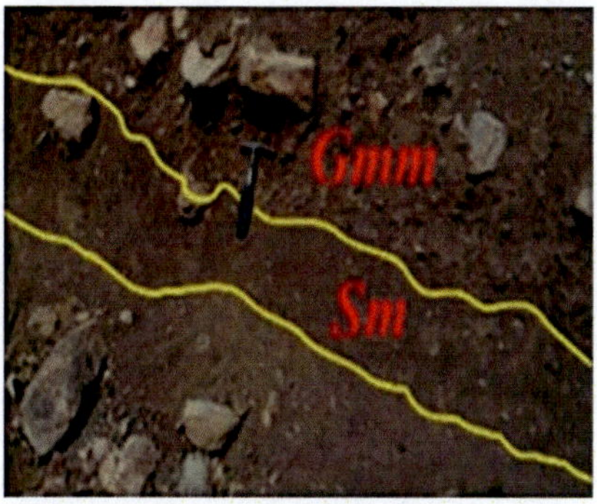

Fig. 3.24 Gcm and Sm facies represent the structural element of GB in the west of the Andimeshk region with erosive contact

3.2.2.2 Structural Element of Channels Element Sediments (Channels Element: CH)

This structural element includes facies Gmm, Gcm, Gh, Gp and Sh. Channel deposits have a clear erosion rule. The geometric figure of these sediments is lenticular and their base is concave upwards. The thickness of canal deposits sometimes reaches more than 10 m, but rarely up to more than 15 m. The lower boundary of these facies is usually sudden or erosive (Fig. 3.25).

The width of the canals varies between 1 and 20 m. Most canal fillers have a multi-cycle or multistory nature, and in each cycle or cycle of canal filling, a decrease in grain size is observed upwards. The presence of a coarse-grained conglomerate at the base of the channel-filling sediments indicates a sudden increase in the sediment flow rate at the beginning of each downward cycle.

Coarse sediments, in addition to high energy, show a relatively high sediment supply rate, which is higher than the available space and leads to a high increase in sediment. This structural element is abundant in the studied outcrops. The large lateral expansion of the canal indicates the high displacement of this sub-environment.

3.2.2.3 Structural Element of Sediment Gravity Flow Deposits (SG)

Sediment gravity flow deposits include Gmm, Gci, Gmg and Gcm sedimentary facies. These structural elements generally has a definite erosion rule and their geometric figures are wide and elongated (Fig. 3.26).

In the study areas, this structural element generally has both sloping and erosive contacts that are covered by gravelly or sandy layered figurers and may also be present in channel deposits.

The frequency of this structural element is higher in the upstream areas of the studied river. The thickness of sediment gravity flow deposits in most cases reaches more than 2 m and eventually up to 10 m. The width of these deposits varies between 7 and 11 m. Sediment gravity flow deposits show deposition of non-cohesive gravity

Fig. 3.25 Facies forming the structural elements of channel-filling sediments with erosion boundary between Gh, Gmm and Gcm facies in the northeast of the Andimeshk region

flow (Miall 2014). The presence of these elements in the canal fillings probably indicates the instability of the canal margins and their collapse.

3.2.2.4 Structural Element of Sandy Layer Forms (Sandy Bedforms: SB)

These structural elements consist of sheet-figured masses of sand deposits (Sh facies). This structural element often has a definite rule and is eroded by the channel. These structural elements are commonly found in the study areas along with the GB element. Sand-layered forms in the study areas are often about half a meter thick and sometimes more than 3 m thick. The width of these masses varies between 4 and 16 m.

In addition, the lower contact is generally sloping (Fig. 3.27). This structural element is probably due to the plastic streams and in areas close to the source with Fig. 3.27 Erosion boundaries and the geometric figure of wide and wide sheets (Miall 2014).

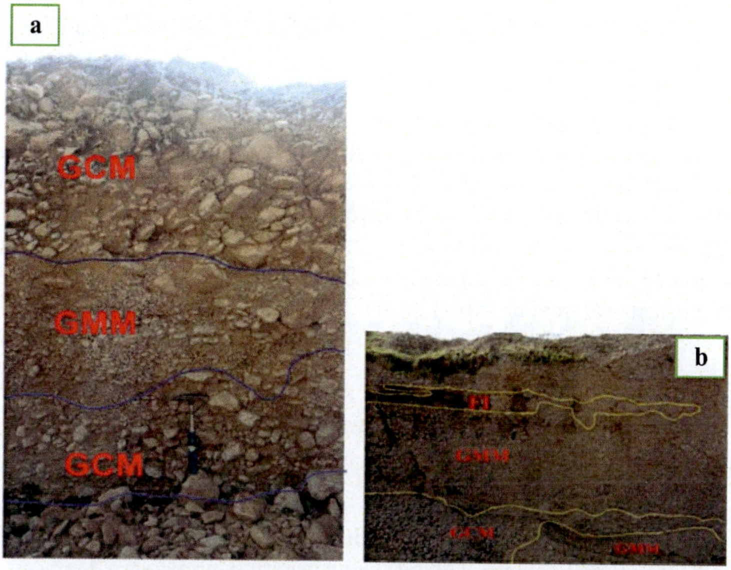

Fig. 3.26 Facies forming the structural elements of SG along **a** the slope and erosion boundary between Gcm and Gmm facies in the north of Dezful city, **b** slope and erosion boundary between Gcm and Gmm facies in the north of Ramhormoz city

Fig. 3.27 Sh facies in the north of Andimeshk city

3.2.2.5 Structural Element of Fine-Grained Off-Channel Sediments (Overbank Fines: FF)

This structural element includes Fm and Fi facies. The geometric figure of these structural elements is laminate or cover-like (Fig. 3.28). In the study areas, especially in the Ramhormoz area, the base of these sediments is clear and the upper surface is usually eroded. The thickness of these layers in these areas varies between 1 and more than 8 m. These facies assemblages are formed by the deposition of suspended flow in the floodplain or in the form of cover sediments on gravel dams, with a geometric figure of sheets (Foix et al. 2013).

3.3 Calculation of Long-Standing Hydrological Parameters with the Help of Sedimentary Properties

Investigation and study of long-term hydrological estimates of river flow using two methods of regime (Regime Method) and capacity (Competence Method) which is based on the estimation of long-standing floods (Sridhar et al. 2013).

The flow intensity method is based on the experimental relations of flow, based on the dimensions of the long channel and the sediment properties by Williams (1984). The capacity method uses the experimental relationships between grain size and the hydrological conditions required to initiate sediment movement (Sridhar et al. 2013).

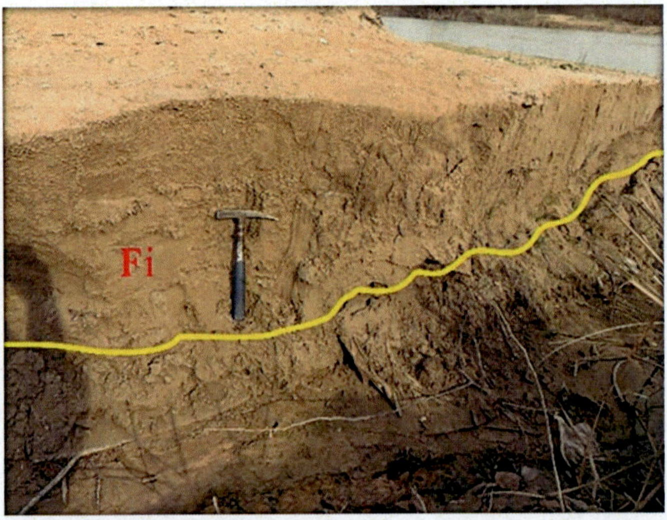

Fig. 3.28 Lateral expansion of Fi massif facies in the southeast of Ramhormoz

In this study, paleo channel dimensions and sedimentological properties have been used to reconstruct the ancient discharge rate. Experimental relation (Williams 1984) has been used to calculate the strength of the river.

In this formula, w is the flow rate per unit area (W m^{-2}) and d is the average grain diameter in meters.

In these studies, the maximum size of gravel parts in each rock facies in 4 study sections (2 sections in Dezful and 2 sections in Ramhormoz) from the upstream, middle and near downstream sections, measured in field studies and an average of 10 pieces Large is used to calculate the current strength (Figs. 3.29 and 3.30).

Costa's (1983) equation is used to calculate the average flow velocity based on the size of gravel parts

$$V = 0.18d^{0.49}$$

In this equation, v is the water velocity in terms of m s^{-1} and d is the mean axis length and the largest part of the gravel.

In these studies, to determine the average value of channel depth from the thickness of diagonal layers in Gravelly facies using Allen equation (1968) has been calculated

$$h = 0.086(dm)^{1.19}$$

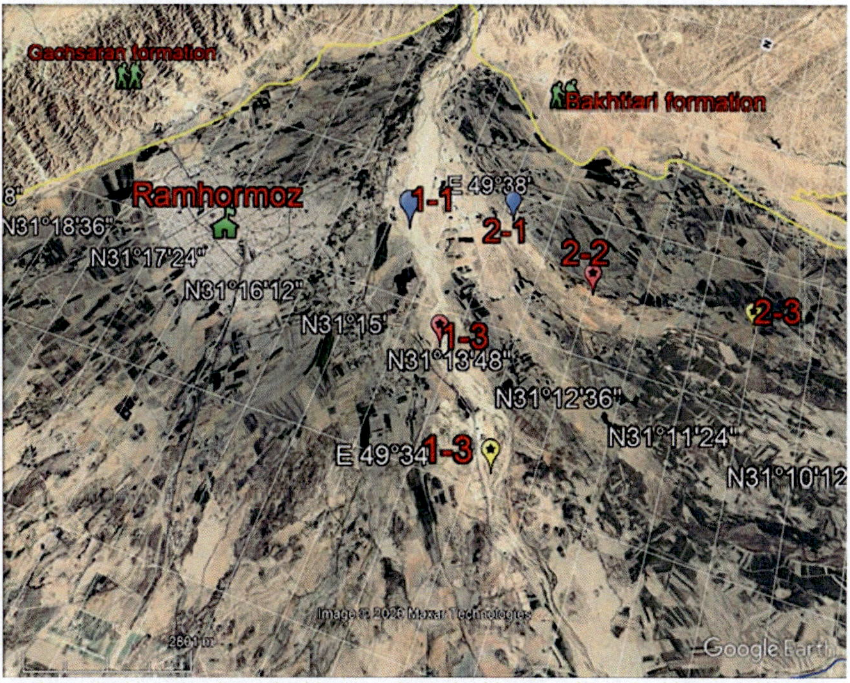

Fig. 3.29 Location of sections No. 1 and 2 in Ramhormoz study area in southwestern Iran

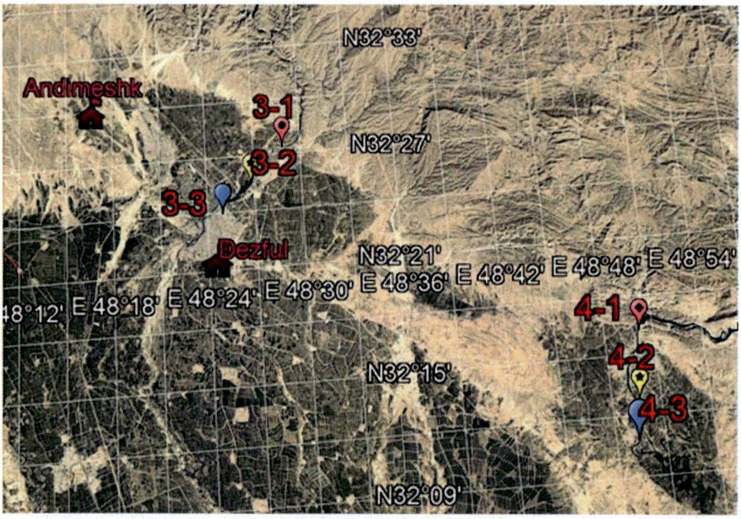

Fig. 3.30 Location of sections No. 1 and 2 in the Dezful study area in southwestern Iran

In this formula, h is the average thickness of the diagonal layers in meters and dm is the average depth of the channel in meters.

Also, the average and maximum value or peak discharge current is estimated using William's 1984 equations according to the following formulas:

$$Q = 0.29\mathrm{Wb}^{1.28}\mathrm{D\ max}^{1.10}$$

$$Q_{2.33} = 2.66\mathrm{Wb}^{0.90}\mathrm{D\ max}^{0.68}$$

In these equations, Q is the average discharge current in $m^3\ s^{-1}$, $Q_{2.33}$ is the average annual peak discharge peak in $m^3\ s^{-1}$, D max is the maximum channel depth in meters and Wb is the channel width in meters.

3.3.1 Estimate the Current Strength Using the Part Size

According to the studies, the maximum size of the measured parts in the study areas is related to the two facies: Gcm and Gmm. The size of parts in Gmm facies in the Dezful region reaches up to 30 cm and in the Ramhormoz region up to 22 cm.

The size of parts in Gcm facies in the Dezful region is a maximum of 24 cm and in the Ramhormoz region, it is 15 cm.

According to the values calculated for river strength based on the size of the above sections in the four locations studied, the maximum estimated value for river strength is related to Gmm rock facies in section 1 in the northeast of Ramhormoz city, which may be related to a major flood event. Calculated values for river strength increase from the base to the top of the section, which indicates a gradual decrease in sediment up to the top of the section.

The presence of finer grain channel facies (Sh facies) in the Dezful region indicates periods with more sedimentation, which can be due to strong flood currents in different climatic periods. Flow velocity values are calculated based on part size according to Costa's formula (1983) from 0.56 to 2.62 m s^{-1}, which varies from 0.61 to 2.62 m s^{-1} in the Ramhormoz region, while in the Dezful region, this range varies between 0.56 and 1.67 m s^{-1}.

As can be seen, the maximum current velocity that belongs to the Ramhormoz region corresponds to the maximum current power (591.82 W) and the maximum current discharge rate (166.32 m^3 s^{-1}). According to the obtained data, Gmm and Gcm facies in both studied areas are associated with high values of power and flow velocity, while Gh and Gp facies in both regions have the lowest values of calculated velocity and power.

Vertical changes in rock facies as well as changes in the thickness of oblique floors and the size of the oscillating sections suggest hydrological conditions (Table 3.6).

.

3.3.2 *The Thickness of the Category of Diagonal Floors and Estimating the Amount of Long-Term Drainage*

A study of the thickness of the diagonal floor to calculate the maximum depth of the channel using the equation presented by Allen (1968) shows that in the study areas the thickness of the diagonal floor varies between 1.14 and 2.21 m and the calculated depth of the channel is between 1.12 and 2.95 (1.12–1.36 for Dezful region and 2.36–2.95 for Ramhormoz region).

Using the Williams (1984) equation, the width of the canal is calculated and its value varies between 19.16 and 66.3 (19.16–21.12 for the Dezful region and 25.12–66.3 m for the Ramhormoz region). Similarly, the average and maximum annual discharge is calculated based on Williams's (1984) equations.

The average discharge varies between 15.36 and 78.12 (15.36–29.15 in the Dezful region and 68.36–78.12 in the Ramhormoz region). As can be seen, the maximum annual discharge, which corresponds to the maximum thickness of the oblique groups, belongs to the Ramhormoz area (203.21 m^3 s^{-1}).

In addition, the minimum annual discharge history in the study areas belongs to the Dezful region, which is consistent with the minimum thickness of the oblique strata (56.32 m^3 s^{-1}) (Table 3.7).

According to studies, this data is largely consistent with the statistical data of water discharge of Khuzestan Water and Electricity Organization.

Based on these data, the annual water discharge in the 50-year period in section No. 1 of the Ramhormoz region is about 187.67–218.31 m^3 s^{-1} and in section No. 2

Table 3.6 Estimation of long-term discharge strength and velocity based on gravel particle size in different sedimentary facies of the studied areas in sections 1, 2, 3 and 4

Current discharge rate ($m^3\ s^{-1}$)	Channel cross section A (m^2)	The power W (wm^{-2})	Speed V ($m\ s^{-1}$)	Maximum average grain size d (mm)	Sedimentary facies	Coordinates	Location	Cutting number
166.32	70.21	591.82	2.62	220.8	Gmm	31° 14′ 45″ N 49° 36′ 59″ E	Ramhormoz	1–1
108.28		113.81	1.31	112.3	Gmg	31° 14′ 2″ N 49° 35′ 54″ E	Ramhormoz	1–2
156.82		526.25	2.36	239.8	Gcm	31° 12′ 39″ N 49° 33′ 50″ E	Ramhormoz	1–3
31.6	65.3	61.36	0.95	248.2	Gmm	31° 13′ 39″ N 49° 37′ 34″ E	Ramhormoz	2–1
66.4		438.25	2.16	148.2	Gcm	31° 12′ 33″ N 49° 36′ 52″ E	Ramhormoz	2–2
17.92		16.2	0.61	110.6	Gp	31° 10′ 48″ N 49° 37′ 1″ E	Ramhormoz	2–3
23.9	153	39.12	0.59	109.4	Gh	32° 27′ 6″ N 48° 28′ 40″ E	Dezful	3–1
68.3		263.21	1.67	96.4	Gp	32° 24′ 46″ N 48° 26′ 03″ E	Dezful	3–2
81.2		16.8	1.12	231.3	Gcm	32° 23′ 57″ N 48° 24′ 04″ E	Dezful	3–3
86.35	36.2	32.12	1.06	180.1	Gmg	32° 16′ 14″ N 48° 50′ 09″ E	Dezful	4–1
78.12		14.6	0.58	10.1	Sh	32° 14′ 09″ N 48° 49′ 18″ E	Dezful	4–2

(continued)

Table 3.6 (continued)

Current discharge rate ($m^3 s^{-1}$)	Channel cross section A (m^2)	The power W (wm^{-2})	Speed V ($m s^{-1}$)	Maximum average grain size d (mm)	Sedimentary facies	Coordinates	Location	Cutting number
79.26		15.3	0.56	92.9	Gp	32° 11′ 57″ N 48° 49′ 30″ E	Dezful	4–3

Table 3.7 Estimation of channel dimensions and maximum flow for gravelly layer figures based on the thickness of the diagonal strata handle (based on Williams relation, 1984)

Annual flow discharge history Q2.33 ($m^3 s^{-1}$)	Medium discharge current Q ($m^3 s^{-1}$)	Channel width w (m)	Maximum channel depth dm (m)	The thickness of the handle of the diagonal layers h (m)	Location	Cutting number
203.21	68.36	66.3	2.36	0.26	Ramhormoz	1
66.36	78.12	25.12	2.95	0.21	Ramhormoz	2
56.32	29.15	21.12	1.36	0.14	Dezful	3
106.23	15.36	19.16	1.12	0.16	Dezful	4

of the Ramhormoz region 49.27–71.49 $m^3 s^{-1}$ and in section. The number one area of Dezful is 42.29.96–68.32 $m^3 s^{-1}$ and the number two cut of Ramhormoz area is 78.92–106.37 $m^3 s^{-1}$.

Estimation of long-term discharge values based on sediment characteristics shows that hydrological conditions during the deposition of gravels can be related to climate change, including rainfall in different seasons in these areas.

The maximum river power and the maximum annual discharge are compatible with the sedimentary gravity flow (element) deposits (SG) and the lowest flow power and annual discharge are related to the dam structural element and the gravel layer figures (GB element).

3.4 Classification of Alluvial Fans

There are several methods for classifying alluvial fans based on their figure, texture, morphology, size and other physical properties. The best classification that can be introduced for these terrain features is a classification based on all the properties mentioned so that it can have the best description about them.

The following section provides an example of this division for study areas in southwestern Iran to become more familiar with this type of division.

According to the studies, including granulometric and facies information, alluvial fans of the Dezful and Ramhormoz regions can be divided into different sections from the top to the bottom, which will be studied in the following section.

3.4.1 Ramhormoz Alluvial Fan Division (in Southwestern Iran)

In the Ramhormoz alluvial fans, due to the large distance between the apex and the downstream, the sediments are well separated based on the particle size, and as a result, the sediments are divided into three categories: coarse grain (proximal part), medium grain. (Medial part) and fine-grained (distal). In this classification, in addition to particle size, the slope of the environment, morphotectonic properties and geomorphology of the region are also considered. The following section examines the properties of the proximal, medial and distal sections of this alluvial fan.

3.4.1.1 Ramhormoz Alluvial Fan Proximal Section

Sediments of the proximal part (close to the source), which mainly consist of coarse-grained sediment particles, extend from the apex of the alluvial fan to the middle parts of the alluvial fan (Fig. 3.31).

This part is more protruding in the north, northwest and in front of Ramhormoz drift and includes alluvial deposits that are the result of erosion of Gachsaran Formation outcrops in Ramhormoz drift wall and generally have relatively weak sorting (Fig. 3.32a, b).

In addition to the Gachsaran Formation, many detrital particles and fragments enter the basin from the upstream parts, including the Mal-Agha River (Izeh city in southwestern Iran), which mainly belong to the Aghajari and Bakhtiari Formations.

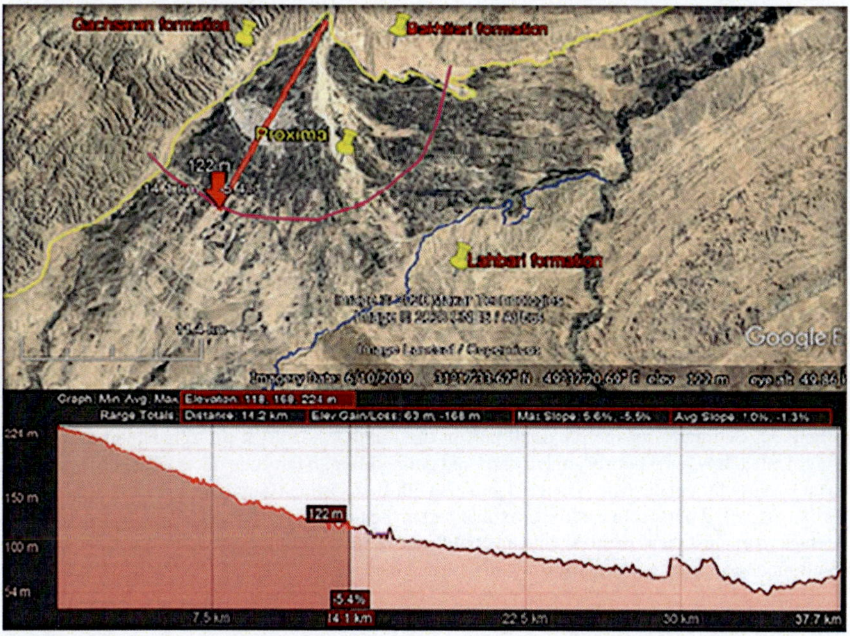

Fig. 3.31 Position and range of proximal section in Ramhormoz large alluvial fan cone in satellite images

Fig. 3.32 Sediments of the proximal part in the Ramhormoz area. **a** Degradable sediments and fragmented flow, with poorly angled parts and poor sorting, **b** fan cone ducts with poorly supported backwash sediment deposits, **c** angled gravels with Bad sorting (Gmm) with the erosive rule in the proximal part, **d** perspective of the proximal part at the foot of Gachsaran Formation—west of Shifa village, northeast view, **e** gravel with an abundance of mud particles caused by debris flows, **f** parts with moderate circulation and sorting due to river currents

In fact, this part, which includes units close to the top of the alluvial fan, is composed of gypsum gravel grains with sand and sometimes deposits of fine silt and clay grains (Fig. 3.32c, d).

The grain size decreases from the top of the alluvial fan downstream due to the decrease in flow intensity and gradually becomes the deposits of the medial part. In vertical cutting, the grain size decreases from the base upwards, and on the surface, it often contains fine silt grain deposits and gypsum clay.

These deposits are usually cut by deep waterways and in some places have the appearance of a satellite hill. In addition to gypsum grains, calcareous and sand gravels can be seen in this section, which have less roundness than the expressed gypsum gravel grains (Fig. 3.32e, f).

3.4.1.2 Medial (Middle) Part of Ramhormoz Alluvial Fan

This unit, which protrudes in the middle part of the study area and in the forehead of the Ramhormoz drift, includes alluvial deposits that have been formed as a result of erosion of Gachsaran Formation outcrops in the Ramhormoz drift wall (Fig. 3.33).

The sediments of this section, which form the middle part of the alluvial fan, mainly contain medium-grained particles and with less frequency include coarse-grained and fine-grained particles (Fig. 3.34).

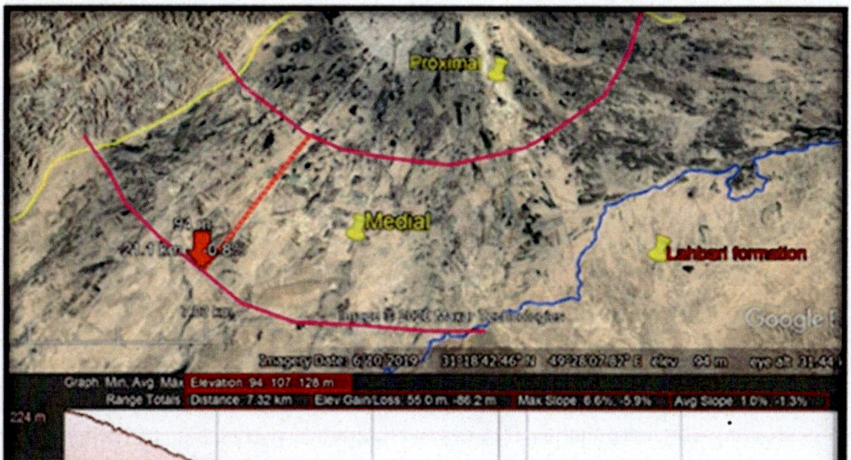

Fig. 3.33 Position and range of the medial part in the large Ramhormoz alluvial fan in satellite images

Fig. 3.34 Sediments of the medial part in Ramhormoz area. **a** Close view of the medial part of the Ramhormoz alluvial fan, **b** far view on the southwestern edge of the anticline, **c** close view of the medial part, **d** medial part on the southwestern edge of the anticline, look to the southwest

These sediments are composed of sand and clay particles along with gypsum gravel grains that have been deposited at a longer distance from the source due to the intensity of the flow and changes in the slope of the area.

3.4.1.3 Distal Part of Ramhormoz Alluvial Fan

The sediments of this section mainly consist of fine-grained silty and clay deposits with sandy and gypsum lens protrusions that form the end parts of the Ramhormoz alluvial fan and gradually turn into floodplain sediments (Fig. 3.35). Sediments of this section, which are mainly free of coarse-grained particles due to reduced flow intensity, are not observed in the southern part of the study area due to interruption by Lahbari drift, but in the southeastern and southwestern parts of the region to the flood plain. Continue (Fig. 3.36; Table 3.8).

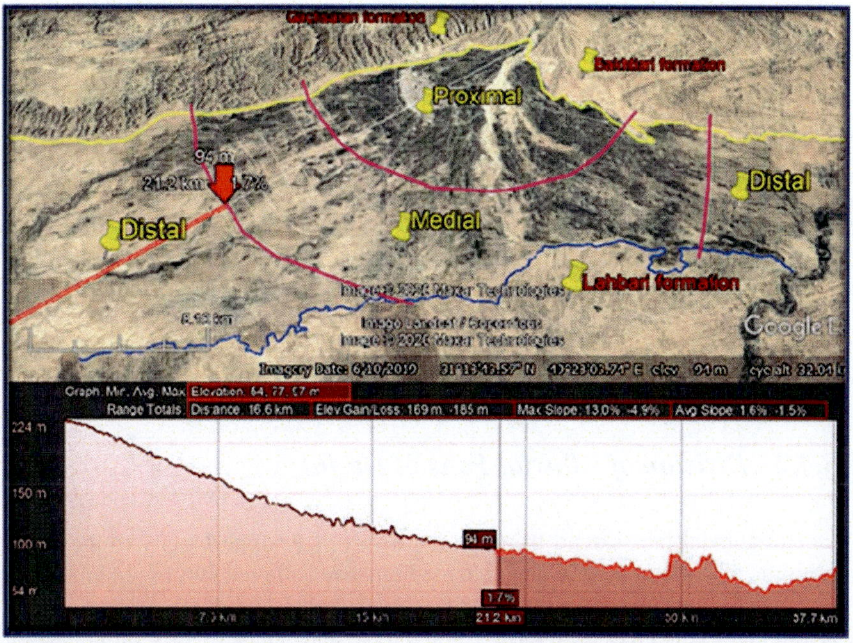

Fig. 3.35 Position and range of the distal part in Ramhormoz large alluvial fan in satellite images

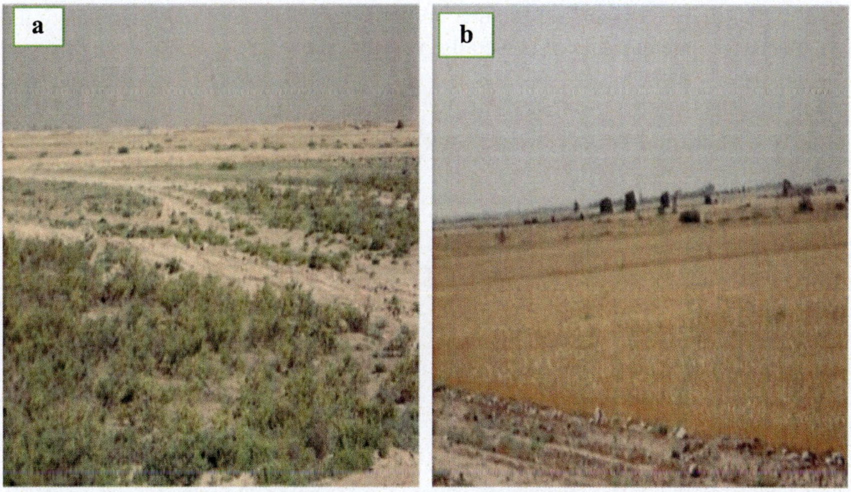

Fig. 3.36 Perspective of the distal part in Ramhormoz plain, **a** southeast of Ramhormoz, **b** southwest of Ramhormoz

Table 3.8 Abstract of general sedimentary properties of different parts of Ramhormoz alluvial fan in the study sections

Part	Name	Sorting	Form	Roundness	Sphericity	Maturity	Matrix	Source
Proximal	G.S.M	Medium to poor	L›I›S	Semi-angled	Semi stretched	Immature	Silt, clay, medium sand	BK-Gs
Media	S.G.M	Medium	L›I›S	Semi-angular	Semi stretched	Submature	Silt, clay, fin sand	BK-Gs
Distal	M.S.G	Medium to good	L=I›	semi stretched	Semi stretched	Submature	Silt, clay, fin sand	BK-Gs

3.4.2 *Division of Alluvial Fans of Dezful*

Dezful alluvial fans extend from the northwest of Andimeshk city to the north of Shushtar city and include 37 small to medium-sized alluvial fans (5–14 km long and less than 5 km wide) originating from Bakhtiari, Aghajari and Lahbari sections. They are similar to the great Ramhormoz alluvial fan and can be divided into three parts: proximal, medial and distal. In dividing these alluvial fans, it is very important to study the role of slope because due to the small size and expansion of these alluvial fans, it is difficult to separate the medial parts from the distal only by changing the size of sediments in many of these alluvial fans.

3.4.2.1 Proximal Part (Close to the Origin) of Dezful Alluvial Fans

Sediments of this section in different alluvial fans have been studied in the Dezful region due to similar morphotectonic and geomorphological properties of these alluvial fans which have similar sedimentary properties. In addition, in all the alluvial fans studied, due to their similar origin (Aghajari and Bakhtiari Formations), their sedimentary and rock properties are very similar. The sediments of this section in the studied alluvial fans include coarse-grained gravels of limestone type with medium to relatively good sorting and rounding, along with medium-grained sand particles and fine-grained grain particles. These sediments are mainly formed by retail or debris flows and the presence of a hard source (Bakhtiari Formation) prevents the large deformation of these sediments. Figure 3.37 shows the position of the proximal section in satellite images and Fig. 3.38 shows the perspective of one of the alluvial fans in the Dezful region.

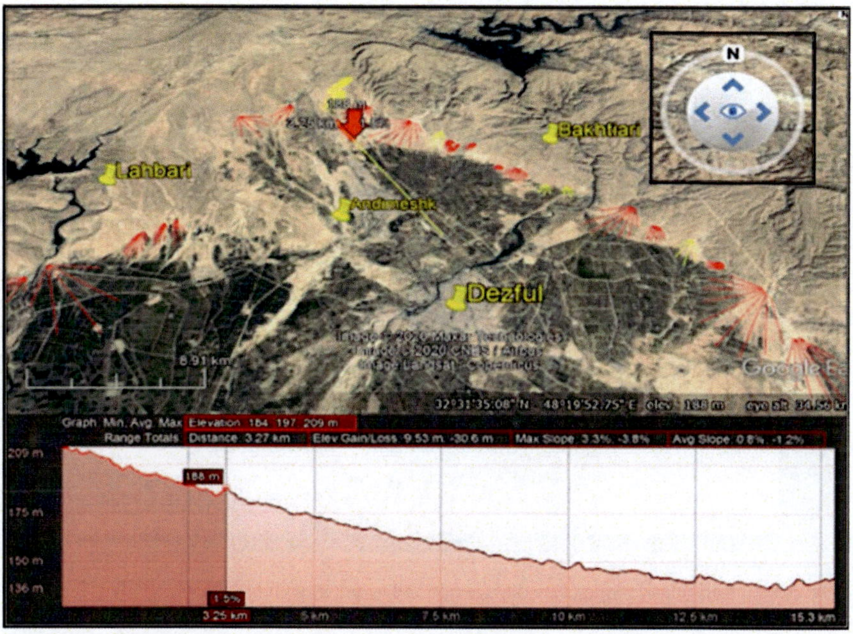

Fig. 3.37 Position and range of proximal section in Dezful alluvial fans in satellite images

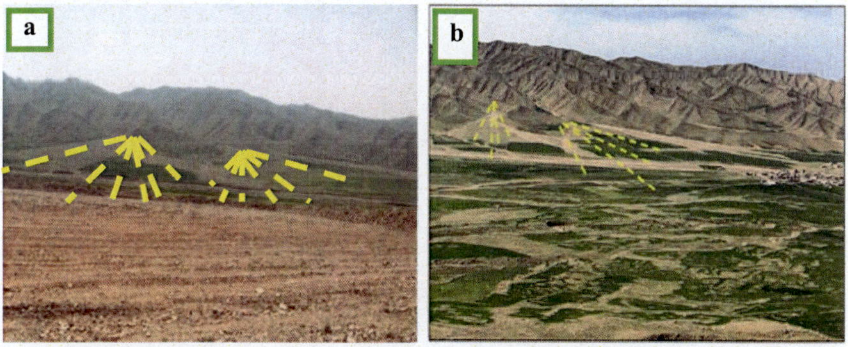

Fig. 3.38 Proximal section view in Dezful study area. **a** Northeast of Andimeshk, **b** North of Dezful

3.4.2.2 Medial Section of Alluvial Fans of Dezful Region

This part, which mainly consists of medium grain parts with less frequency of coarse and fine grain parts, is in the middle parts of alluvial fans of the Dezful region (Fig. 3.39). Particle sorting in this section is relatively weak to medium and the particles are semi-rounded. From Andimeshk to Shushtar, the area of this section is reduced and particle sorting is reduced (Figs. 3.40 and 3.41).

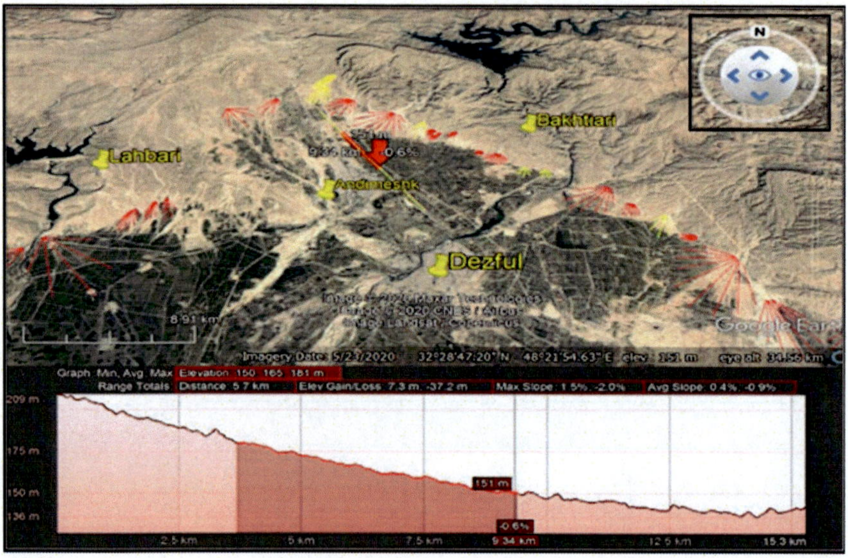

Fig. 3.39 Position and range of medial section in Dezful alluvial fan cone in satellite images

Fig. 3.40 Near view of sediments in the medial part in the north of Andimeshk. **a** Northwest of Andimeshk, **b** North of Dezful, southwest of Iran

3.4.2.3 Distal Section in Dezful Alluvial Fan

These sediments include the cities of Andimeshk and Dezful and many fertile agricultural lands, so accurate knowledge is very important, especially in the economic sector and urban hazards (Fig. 3.42). In addition, the sediments of this section, which

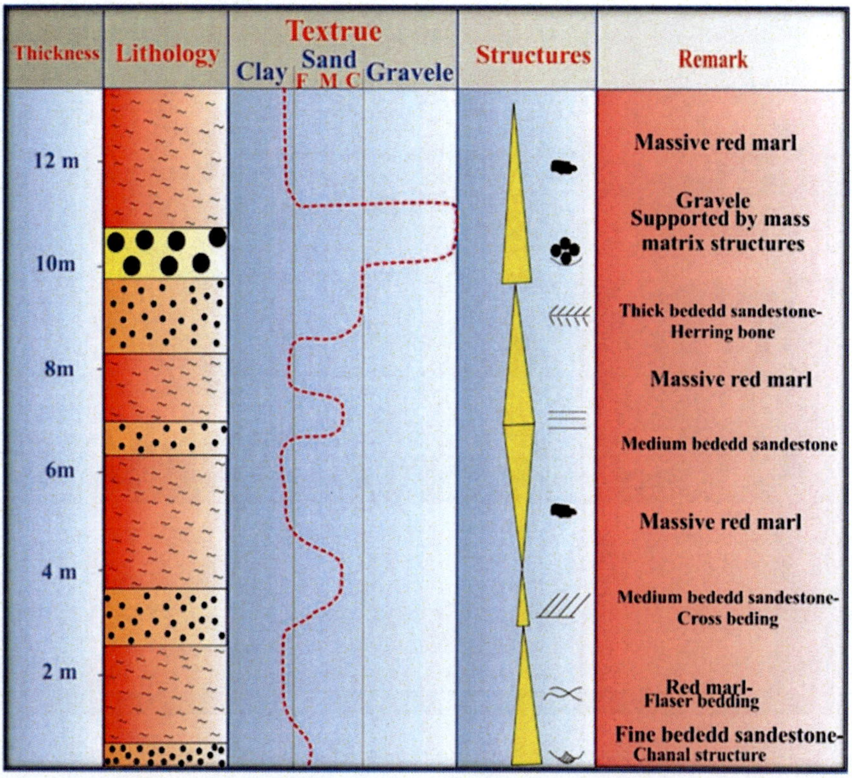

Fig. 3.41 The stratigraphic column of the medial section in the Dezful study area with the alternation of fine, medium and coarse lithologies with upward cycling cycles

have the highest frequency and expansion compared to other sections of alluvial fans, include fine-grained silt and clay particles, which are darker in color than the distal section of the Shushtar alluvial fan and relatively have bumps. They are seen better in Fig. 3.43 (Table 3.9).

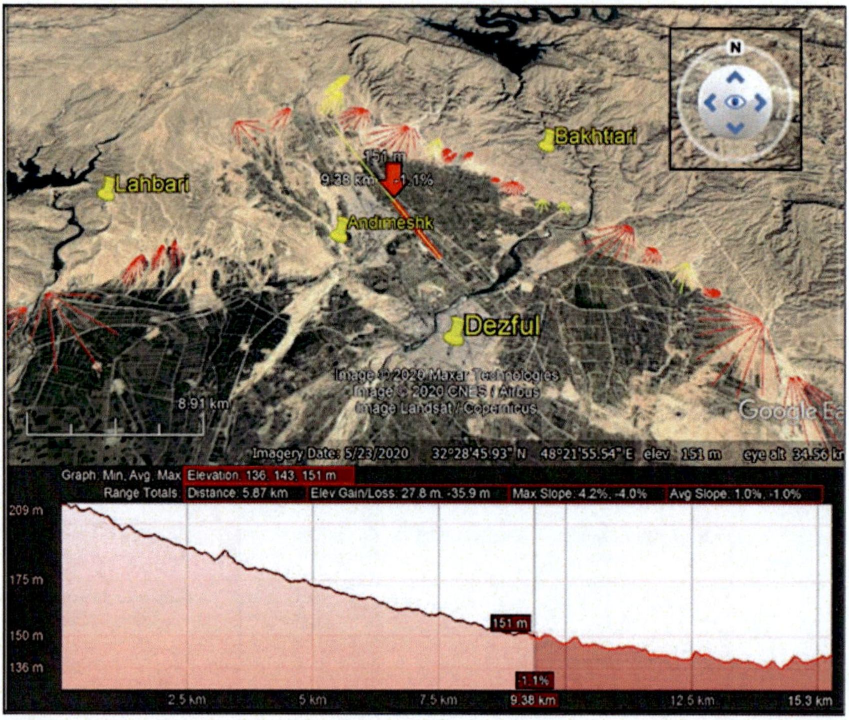

Fig. 3.42 Position and range of the distal part in the Dezful alluvial fan cone in satellite images

Fig. 3.43 Close view of sediments in the distal part of the alluvial fans of West Andimeshk, southwest of Iran

Table 3.9 Abstract sedimentary properties of different parts of Dezful alluvial fan in the studied sections

Part	Name	Sorting	Form	Roundness	Sphericity	Maturity	Matrix	Source
Proxima	G.S.M	Well to Medium sorting	L›I›S	Semi-rounded	Semi stretched	Submature	Clay, silt, medium sand	Aj, Bk, La
Medial	S.G.M	Medium sorted	L›I›S	Semi-rounded	Semi stretched	Immature	Clay, silt, fin sand	Aj, Bk, La
Distal	M.S.G	Sorted	L=I›S	Rounded	Medium	Immature	Clay, silt, fin sand	Aj, Bk, La

References

Abu M (2018) Sedimentary facies and depositional environments of the neoproterozoic sediments of the Gambaga-Nakpanduri massifs, Voltaian Basin. J Geol Min Res 10(5):48–56
Abu M, Sunkari ED, Gurel G (2020) A. Paleocurrent analysis, petrographic, geochemical and statistical appraisal of Neoproterozoic siliciclastic sediments, NE Voltaian Basin, Ghana: a multidisciplinary approach to paleogeographic reconstruction. J Sediment Environ (z) (online)
Allen JRL (1968) The nature and origin of bed-form hierarchies. Sedimentology 10:161–182
Apolinarska K, Pleskot K, Pełechata A (2020) The recent deposition of laminated sediments in highly eutrophic Lake Kierskie, western Poland: 1-year pilot study of limnological monitoring and sediment traps. J Paleolimnol 63:283–304
Arzani N (2004) Sediments of alluvial fans and plain (Fluvial Fans) and their role in controlling the discharge of aqueducts from Abarkuh plain (west of Yazd province). In: National conference on aqueducts, Gonabad, Iran
Barkashani H, Block I, Mahmoudi Qaraei H, Mahboobi A, Mousavi Harami R (2014) Rock facies and interpretation of tectonic position of the source area of Miocene detrital siliceous sediments using petrographic and geochemical data in Binalood area, north of Neishabour. Sediment Facies 7(1):19–32
Blair TC, McPherson JG (1999) Grain size and textural classification of coarse sedimentary particles. J Sediment Res 69:6–19
Boggs SJ (2018) Principles of sedimentology and stratigraphy, 8th edn. Pearson Education, Upper Saddle River, New Jersey, pp 662–697
Chen A, Ng Y, Zhang E, Tian M (2020) Dictionary of geotourism. Springer, Singapore (online)
Costa JE (1983) Palaeohydraulic reconstruction of flash-flood peaks from boulder deposits in the Colorado Front Range. Bull Geol Soc Am 94:986–1004
Eizenhofer PR (2020) Regional geology across the Solonker Suture Zone. In: Subduction and closure of the Palaeo-Asian Ocean along the Solonker Suture Zone: constraints from an integrated sedimentary provenance analysis. Springer, Singapore (online)
Eoff JD (2014) Sedimentary facies of the upper Cambrian (Furongian; Jiangshanian and Sunwaptan) Tunnel City Group, Upper Mississippi Valley: newinsight on the old stormy debate. Sed Geol 302:102–121
Fayazi Borujeni M, Mortazavi Mehrizi M, Khaneh Bad M (2017) Investigation of sedimentary facies and sediment deposition pattern in Molid river (southeast of Ghaen). Strat Sedimentol Res 33(1):101–128
Foix N, Paredes MI, Giacosa RE (2013) Fluvial architecture variations linked to changes in accommodation space: Río Chico Formation (Late Paleocene), Golfo San Jorge basin, Argentina. Sediment Geol 294:342–355
Folk E (1980) Petrography of sedimentary rocks. Hemphill Publishing Company, 182 pp
Ghorbani M (2019) Lithostratigraphy of Iran. Springer Nature, Switzerland, pp 296–328
Hernandez-Hinojosa V, Montiel-Garcia PC, Armstrong-Altrin JS, Nagarajan R, Kasper-Zubillaga JJ (2019) Textural and geochemical characteristics of beach sands along the western Gulf of Mexico, Mexico. Carpath J Earth Environ Sci 13(1):161–174
Hosseini Barzi M, Bayet Goll A (2010) Facies analysis and sedimentary environment of detrital silicon-carbonate mixtures of shirgasht formation in Kalmard Block, Central Iran. Sediment Facies 2(1):18–21
Karimi R (2014) Easy guide to statistical analysis with SPSS. Hengam Publications, 318 pp
Li K, Deng Q, Hou M (2020) Geochronology and sedimentology of the Huashan Group in the northern Yangtze Block: implications for the initial breakup of the South China. J Earth Sci 167–179
Miall AD (2014) Fluviall depositional systems. Springer International Publication, 316 pp
Mitra R, Chakrabarti G, Shome D (2020) Sedimentation history and depositional model of Palaeo-Mesoproterozoic Tadpatri Formation, Cuddapah Basin, India. J Sediment Environ 5:87–100
Morales JA (2019) Future perspectives of the Spanish Coast. Morales J 167–192

Pourmorad S (2018) Sedimentary mineral resources exploration. Danaeshyaran Publication, Iran, Teharn, p 385

Pourmorad S, Harami R, Solgi A, Ale Ali M (2021) Sedimentological, geochemical and hydrogeochemichal studies of alluvial fans for mineral environment purposes. Lithol Miner Resour 56:89–112

Roy S, Banerjee S (2002) Facies and petrography of the Porcellanite Formation around Chopan, Uttar Pradesh. J Indian Sedimentol 20:195–205

Sarraj RHA, Mohialdeen IMJ (2020) Sedimentology, palynofacies and hydrocarbon generation potential of the Cretaceous Balambo Formation from Zagros Fold-Thrust Belt, Kurdistan, NE Iraq. Arab J Geosci 13:552

Selley RC (1996) Ancient sedimentary environments and their subsurface diagnosis. Chapman and Hall, London, p 300

Sridhar A, Chamyal LS, Bhattacharjee F, Singhvi AK (2013) Early Holocene fluvial activity from the sedimentology and palaeohydrology of gravel terrace in the semi-arid Mahi River Basin, India. J Asian Earth Sci 66:240–248

Stow DAV (2005) Sedimentary rocks in the field: a colour guide. Manson Publishing, London

Tuker ME (2001) Sedimentary petrology, 3rd edn. Blackwell, Oxford, p 260

Williams GP (1984) Palaeohydrologic equations for rivers. Developments and applications of geomorphology. Springer, Berlin, pp 343–367

Chapter 4
Petrographic and Geochemical Studies

Abstract Today, many studies are carried out on the geochemistry of detrital rocks to determine the origin of sediments, which can be used as fundamental studies on the parent rock, tectonic site, climate and chemical weathering (Boggs 2018). Before any geochemical studies, first of all, complete information about the petrographic properties of the studied samples should be obtained and the results of both petrographic and geochemical studies should be verified with each other. In this chapter, it is tried to first conduct a complete study of the petrographic properties of the studied samples in the southwestern region of Iran to adapt to geochemical data and then use the study of major and minor elements in Quaternary sediments in the Ramhormoz region. And Dezful in Khuzestan province, Mansha rock, tectonic position, climate and weathering of Mansha region and groundwater geochemistry of these areas should be studied. This chapter first examines the petrographic results and then the geochemical results and finally interprets the relevant results. This chapter, similar to previous chapters, has practically studied the alluvial fans of southwestern Iran to be considered as a model for the study of alluvial fans all over the world.

Keywords Petrography · Geochemical studies · Mineralogy studies · Oxides · Sub-elements

4.1 Petrography

Petrographic studies provide specific information about sedimentary properties, origin, components and other sedimentary properties of the studied samples (Jalilian 2018). In order to match the geochemical data with the petrographic results, in this section, the aim is to first study the petrographic properties of the upstream cone formations of the studied alluvium and then to study these properties in the studied samples and determine their relationship with upstream formations are discussed.

S. Pourmorad and A. Mohanty, *Alluvial Fans in Southern Iran*, Advances in Geographical and Environmental Sciences, https://doi.org/10.1007/978-981-19-2045-5_4

4.1.1 Petrographic Results of the Studied Areas

The study of petrographic properties of the studied samples in the Dezful and Ramhormoz regions shows that the mentioned samples in large to small sizes originate from Bakhtiari, Aghajari, Gachsaran and Lahbari upstream formations, which are described in the following section. They are examined.

4.1.1.1 Petrography of Upstream Formations in the Dezful Region

Petrographic studies of Bakhtiari Formation samples in the Dezful region show that these samples are mainly composed of limestone fragments (84%) and to a lesser extent quartz monocrystalline and polycrystalline quartz (13%) and feldspar (2.4%) (Table 4.1). The size of the gravels varies from granule to pebble, and all studied specimens are observed as semi-circular to semi-angled with medium sphericity. These conglomerates have a coarse-grained sand matrix with point, linear and convex-concave contacts (Fig. 4.1a). In the studied samples, potassium-type feldspars with moderate weathering are sometimes observed (Fig. 4.1b).

The petrographic study of the Aghajari Formations and Lahbari section in this region shows that in both mentioned formations, rock fragments contain the highest percentage of constituents (average of 84% for the Aghajari Formation and 83% for the Lahbari section). Also, in these two formations, after calcareous shales, monocrystalline and polycrystalline quartz have an average frequency of 14% for the Aghajari Formation and 15% for the Lahbari section. The amount of feldspar in the studied samples of these two formations is about 2.1–2.4% (Table 4.1).

In Aghajari Formation and Lahbari section, detrital siliceous sediments have been identified as petrocytes of lithic arnite. Petrofacis lit arnite is composed mainly of particulate matter (limestone, sediment, sand and volcanic particles), monocrystalline and polycrystalline quartz, feldspar (including plagioclase and potassium) and heavy minerals (such as sphene, tourmaline and opidate). In these two sedimentary units, the semi-angular to semi-rounded constituent particles are point, linear and convex-concave contacts (Fig. 4.1c, d).

In these two units, quartz is generally amorphous, monocrystalline and to a lesser extent polycrystalline. Particles in these two sedimentary units are diverse and include limestone, sediment, sandstone and igneous particles. The most abundant particulate matter in these samples is calcareous aggregates with an average of 69% and less aggregates belong to igneous aggregates (less than 1%). In addition, the quartz is rounded, well assorted, and due to the presence of a clay matrix, it is sub-machined (Fig. 9.3E).

In Bakhtiari, Aghajari and Labri formations, in some samples, heavy minerals such as sphene and tourmaline are scattered, the frequency of which is generally less than 1%. It should be noted that despite the use of heavy minerals in the interpretation of the source because these minerals show different behaviors toward hydrodynamic conditions and geochemical effects, they usually have very variable dispersion in

Table 4.1 Results of petrographic studies in Dezful study area in southwestern Iran

D10	D9	D8	D7	D6	D5	D4	D3	D2	D1	Lbm	Aj	BK	Components of samples		
11.2	7.3	6.4	8.1	8.2	6.7	7.4	8.1	6.3	8.2	9.7	9.1	7.4	Monocrystalline (Qm)		Quartz%
3.5	5.6	4.2	6.6	4.9	5.8	7.2	5.7	5.9	6.5	3.2	3.7	4.7	Polycrystalline (Qp)		
14.7	12.9	10.6	14.7	13.1	12.5	14.6	13.8	12.2	14.7	12.9	12.8	12.1	Quartz sum (Qt)		
1.2	1.2	1.3	1.2	1.6	1.3	0.9	1.2	1.3	1.1	1.6	1.3	1.3	Potassium (K)		Feldspar%
0.7	1	0.3	0.9	0.8	0.6	0.4	1.2	0.3	0.7	0.6	1.1	1.3	Plagioclase (P)		
1.9	2.2	1.6	2.1	2.4	1.9	1.3	2.4	1.6	1.8	2.2	2.4	2.6	Feldspar sum (F)		
0.7	0.8	0.9	0.7	0.8	0.9	0.7	0.8	0.6	0.9	1.3	1.1	1.3	Igneous shale (Lv)	Sediment rubble	Rubble
4.9	7.2	5.3	5.6	4.4	7.4	5.1	6.7	4.6	4.2	6.8	5.2	4.3	Chert (cht)		
71.4	67.4	74.3	71.3	70.8	68.3	71.6	71.1	74.2	68.7	68.4	72.1	72.9	Calcareous (Lc)		
5.4	8.4	6.1	5.6	7.4	7.7	6.7	5.2	5.7	8.7	7.1	6.4	5.8	Sandstone (Lm)		
81.7	83	85.7	82.5	82.6	83.4	83.4	83	34.5	81.6	82.3	83.7	83	Sum of Sediment rubble (LS)		
82.4	83.8	86.6	83.2	83.4	84.3	84.1	83.8	35.1	82.5	83.6	84.8	84.3	Sum Rubble (Lt)		
1	1.1	1.2	0	1.1	1.3	0	0	1.1	1	1.3	0	1	-		Heavy Minerals
Calcilutite	Calcilutite	Calcilutite	Calcilutite	Calcilutite	Calcilutite	Calcilutite	Calcilutite	Calcilutite	Calcilutite	Calcilutite	Calcilutite	Calcilutite	-		Stone name

Fig. 4.1 Microscopic images of sediment particle composition in upstream formations of the study area. **a** Conglomerate of Bakhtiari Formation with semi-assembled, semi-angled, semi-rounded particles with medium to good sphericity in southwestern Iran (Dezful region), **b** point and linear contacts of Bakhtiari Formation gravels in Dezful region, **c** point contact (green arrow), linear contact (blue arrow) and convex-concave contact (yellow arrow) in litarnite sandstones in Aghajari Formation in the north of Dezful region, **d** chert (red arrow) and calcite cement (blue arrow) with quartz with direct and wave extinction in Aghajari Formation in the north of Andimeshk region, **e** siltstone and chert sediments in conglomerates of Bakhtiari Formation in the north of Andimeshk region, **f** Polycrystalline quartz and coarse quartz crystals in calcite cement in Bakhtiari Formation in the northwest of Ramhormoz study area, **g** Potassium type feldspar (yellow arrow) in the matrix of Bakhtiari conglomerates east of Ramhormoz region, **h** Semi-angular to semi-circular sand grains with medium to weak sorting and containing polycrystalline quartz (red arrow) in Gachsaran Formation in southwestern Iran

rocks (Dickinson 1985). Therefore, in this study, these seeds have not been used to interpret the origin.

4.1.1.2 Petrography of Upstream Formations in the Ramhormoz Region

Gachsaran and Bakhtiari are present in the Ramhormoz region as upstream formations of the Ramhormoz alluvial fan. Bakhtiari Formation in this region has almost similar properties to this formation in the Andimeshk region. The formation similar to this formation in the Andimeshk region is mainly composed of gravel (limestone, chert, sand and volcanic particles) with an average of 86%, monocrystalline and polycrystalline quartz with an average of 10%, feldspar (including plagioclase and potassium) with an average of 1.8% of heavy minerals (such as sphene, tourmaline and epidote) with an average of 1.1%. In this sedimentary unit, the constituent particles are generally observed as semi-angled and the size of the gravels in this formation varies from granule to pebble (Fig. 4.1g, f).

The study of sandstone sections in the Gachsaran Formation to find the origin of the studied samples shows that in these samples, rock fragments contain the most abundant percentage of components (average 89%). Also in this formation, after calcareous gravels, monocrystalline and polycrystalline quartz have an average frequency of 8.1%. The amount of feldspar in the studied samples of this formation is about 1.6% (Table 4.2). The constituent particles in these samples are semi-angular to semi-rounded with medium to weak sorting and contain more than 10% of the matrix, in other words, they are in the Imchor to Sub-Machur stage (Fig. 4.1h).

4.1.2 Petrographic Properties of the Studied Sediment Samples

Petrographic studies of the studied samples in Ramhormoz and Dezful regions show that detrital siliceous sediments in these areas are composed of large to medium sizes with a relatively small amount of fine-grained sediments (according to Blair and McPherson classification 1999). In the following section, the studied samples are described, which are mainly in the size of sand and granules.

Petrographic study of the studied samples shows that the main components of these sediments include particulate matter (chert, lime, sand and igneous aggregates) with an average of 83% for Dezful samples and 85% for Ramhormoz, quartz (monocrystalline and Polycrystalline) with an average of 13.7% in the Dezful region and 10.6% in the Ramhormoz region, feldspar (potassium and plagioclase) with an average of 1.9% in the Dezful region and 2.1% in the Ramhormoz region (Tables 4.1 and 4.2).

Table 4.2 Results of petrographic studies in Ramhormoaz study area in southwestern Iran

R11	R10	R9	R8	R7	R6	R5	R4	R3	R2	R1	GS	BK	Components of samples		
8.4	6.7	5.3	6.1	8.2	7.3	6.4	9.4	10.2	8.3	9.4	5.7	9.3	Monocrystalline (Qm)		Quartz%
1.7	1.6	4.1	3.5	2.5	5.9	6	5.5	3.5	2.3	2	1.6	2.1	Polycrystalline (Qp)		
10.1	8.3	9.4	9.6	10.7	13.2	12.4	14.9	13.7	10.6	11.4	7.3	11.4	Quartz sum (Qt)		
1.3	2.1	1.8	2.1	1.9	0.73	0.86	0.97	1.7	0.94	1.3	1.1	1.2	Potassium (K)		Feldspar%
1.6	1.3	0.9	1.2	1.2	0.57	0.34	0.93	0.7	0.76	0.8	0.8	0.9	Plagioclase (P)		
2.9	3.4	2.7	3.3	3.1	1.3	1.2	1.9	2.4	1.7	2.1	1.9	2.1	Feldspar sum (F)		
0.8	0.4	0.1	0.7	0.6	0.6	0.7	0.2	0.4	0.9	1	0.1	1.1	Igneous shale (Lv)	Sediment rubble	Rubble
6.7	4.2	5.9	4.6	6.2	5.1	7.6	4.1	5.5	6.5	4.1	6.8	10.3	Chert (cht)		
72.4	75.5	73.2	75.2	71.6	70	69.7	69.2	72.2	72.1	73.2	74.4	64	Calcareous (Lc)		
5.9	8.2	7.7	5.2	6.6	8.4	7.4	8.7	5.8	7.2	6.9	8.6	10.2	Sandstone (Lm)		
85	87.9	86.7	85.1	84.4	83.5	84.7	82	83.5	85.7	84.2	89.7	84.3	Sum of Sediment rubble (LS)		
85.8	88.3	86.8	85.8	85	84.1	85.4	82.2	83.9	86.6	85.2	89.8	85.4	Sum Rubble (Lt)		
1.2	0	1.1	1.3	1.2	1.4	1	1	0	1.1	1.3	1	1.1	-		Heavy Minerals
Calcilutite	Calcilutite	Calcilutite	Calcilutite	Calcilutite	Calcilutite	Calcilutite	Calcilutite	Calcilutite	Calcilutite	Calcilutite	Calcilutite	Calcilutite	-		Stone name

Petrofasis litho-arnite has been identified in the mentioned detrital siliceous sediments. Sediments The mentioned sediments in the Dezful region have grains with good sorting and medium to good roundness (Fig. 4.2a). These sediments have relatively fine, semi-angled grains in the apex of the Ramhormoz alluvial fan (Fig. 4.2b). In this petrophysis, the size of the quartz varies from fine sand to coarse sand and often has a straight to wavy extinction. The frequency of this type of curates is somewhat higher in the samples of the Dezful region. Some monocrystalline grains contain a variety of inclusions, often of the zircon and muscovite minerals. In general, in both regions, the amount of monocrystalline quartz to polycrystalline quartz is higher, which long-term transport during the re-sedimentation cycle can be one of the effective factors in this phenomenon (Fig. 4.2c).

In addition, potassium feldspar is more abundant in these samples than plagioclase. Potassium-containing feldspar due to its greater chemical stability and more abundance potassium feldspar is more abundant in plasic igneous rocks than plagioclase.

Based on microscopic evidence, feldspars in the study areas are semi-aerated or aerated (Fig. 4.2d–g). Petrographic studies of the studied samples in the target areas and in accordance with the data obtained from the upstream formations of these areas, especially in terms of the presence of gravel in them, which indicates the properties of their source rock (Mousavi Harami 2004) shows. The results of petrography of Dezful samples in terms of percentage and type of main components are largely consistent with the results of petrographic studies of Bakhtiari, Aghajari and Lahbari formations. Also, based on the studies conducted in the Ramhormoz region, it can be concluded that the data obtained from the study of the samples are in line with the data obtained from Bakhtiari and Gachsaran formations (sandstone section) to a large extent and indicate this issue. These formations are considered to be among the main sources of these sediments.

Based on the percentage of major constituents in all studied specimens (quartz, feldspar and particulate matter), these specimens are in the Little Arnite range in the Folk (1980) classification table (Fig. 4.3). In these studies, Dickinson's table (Dickinson 1985) has been used for the tectonic location of the studied areas. The results mentioned in the section on tectonic studies of the studied areas have been given with the help of geochemical data to match the mentioned data. It should be noted that the percentage of chert frequency is considered only in the folk naming by the Folk method (Folk 1980) at the pole of the rocks, but in the analysis of the origin of siliceous-detrital sediments by Dickinson (1985) at the pole of the quartz complex, (Qt) is considered.

4.2 Geochemical Studies

For these studies, samples were prepared in 3 sections from the Dezful region and 2 sections from the Ramhormoz region under mineralogy (XRD), elemental (XRF) and spectrum analysis. Mass spectrometry (ICP-MS) was included in the Geological

Fig. 4.2 Microscopic images of sediment particle composition in the studied areas. **a** Sediments of the middle part of Dezful alluvial fan with grains with good sorting and medium to good roundness, **b** sediments of the middle part of Ramhormoz alluvial fan with relatively fine and semi-angled sorting grains, **c** monocrystalline quartz (arrow Red) and potassium feldspar (orthoclase) (blue arrow), northeast of Andimeshk, **d** semi-altered potassium feldspar (orthoclase) (blue arrow), west of Dezful, **e** decomposing feldspar (red arrow), Carbonate particle (blue arrow), northeast of Dezful, **f** decomposing feldspar (red arrow), chert particle (blue arrow) and carbonate particle (yellow arrow) northeast of Ramhormoz, **g** potassium-containing feldspar (Microcline) (red arrow) East Ramhormoz

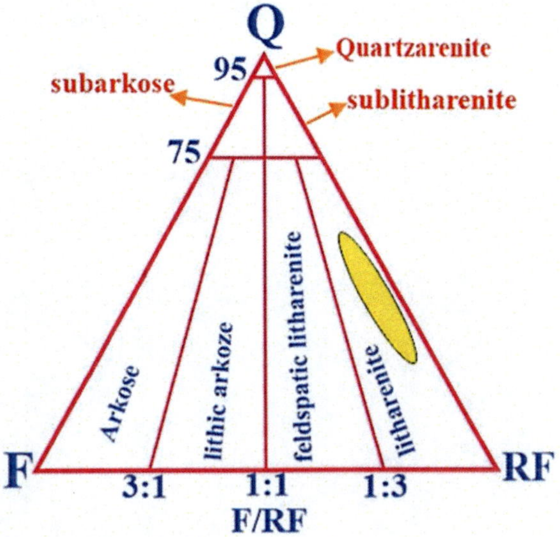

Fig. 4.3 The position of the data obtained from the point count of the studied samples in the Folk (1980) classification model, which shows the Litharenite composition for these samples

Survey of Iran. In addition to these samples, samples from Ramhormoz and Dezful areas under chemical analysis to study the quality of groundwater in the study areas due to their high importance in these areas, especially in the agricultural sector, were studied hydrochemistry, which is described in the relevant section.

4.2.1 Oxides

The results of the analysis of the samples of the Dezful region show that according to these studies, the most abundant oxides in both Dezful and Ramhormoz regions are calcium oxide (CaO) and magnesium oxide and the lowest oxides in these regions are Na_2O, P_2O_5, SrO, TiO_2 and K_2O (Fig. 4.4).The study of the analysis results of these samples in the Dezful region shows that these samples have an average of 31% SiO_2, 7% AL_2O_3, 5% Fe_2O_3, 49% CaO, 3% MgO and the average percentage of Na_2O, P_2O_5, SrO oxides. TiO_2 and K_2O are about 1% each (Table 4.3). In the Ramhormoz region, the studied samples have an average of 28% SiO_2, 5% AL_2O_3, 4% Fe_2O_3, 53% CaO, 4% MgO, 2% SO_3, and the average of other elements is about 1% for each element (Table 4.4).

4.2.2 Sub-Elements

The results of the study of sub-elements in the study areas show that the most abundant sub-element in the study areas is the element titanium (Ti). The average of this

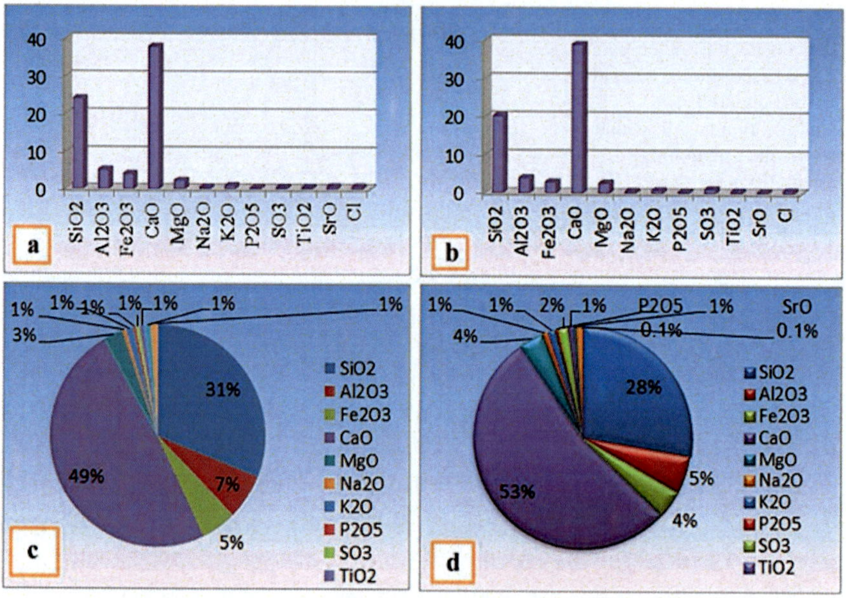

Fig. 4.4 Average percentage of main elements in samples taken from Dezful and Ramhormoz regions. **a**, **b** Rectangular diagrams of the total average percentage of each element before calculating the total percentage in Dezful (**a**) and Ramhormoz (**b**) samples before calculating the total percentage. **c**, **d** Circular diagrams of the total percentage of each element in Dezful (**c**) and Ramhormoz (**d**) harvest samples

element in the Dezful region is 1842 ppm and in the Ramhormoz region is 1451 ppm. In the study areas, after titanium element, other elements in terms of ppm include Zr elements (Dezful 101 and Ramhormoz 72.8 ppm), V (Dezful 57 and Ramhormoz 45.6 ppm), Ce (Dezful 19.1 and Ramhormoz 21.1 ppm) and La (Dezful 12.2 and Ramhormoz 12.4 ppm). Tables 4.5 and 4.6 show the frequency of other sub-elements of Dezful and Ramhormoz regions in terms of PPM.

4.2.3 Rare Earth Elements

Rare earth elements or rare earth metals with the abbreviation "REEs" include a set of 17 elements of the periodic table (Krishnamurthy 2020).

These elements include fifteen elements of lanthanides and two elements of scandium and yttrium (Ghosal et al. 2020). The most abundant rare earth elements in the Dezful region include Ce with an average of 19.26 ppm, followed by La with an average of 12.2 ppm, Nd with an average of 10 ppm, Y with an average of 9.81 ppm, Sc with an average of 5.8 ppm, M, Pr with an average of 2.8 ppm, Sm with an average of 2.24 ppm, Gd with an average of 1.9 ppm and Dg with an average of 1.74 ppm.

Table 4.3 Some results of oxide analysis by XRF in the Dezful region in terms of percentage for training

D23	D22	D21	D20	D19	D18	D17	D16	D15	D14	D13	D12	D11	D10	D9	D8	D7	D6	D5	D4	D3	D2	D1	No
21	20	24	27	20	18	19	23	24	19	20	21	26	17	14	19	19	27	29	28	27	28	23	SiO_2
6.7	4.8	3.2	6.8	5.2	5.8	3.6	3.2	3.7	2.7	2.6	2.8	2.6	2.7	2.5	2.1	3.9	1.9	1.8	2.7	3.1	3.2	2.9	Al_2O_3
5.8	6.1	3.2	4.6	3.7	4.1	3.8	5.2	4.1	2.3	2.1	3.9	4.1	2.7	2.1	2.9	3.2	2.8	2.3	5.9	2.8	3.1	2.6	Fe_2O_3
36.4	31.8	40.1	31.4	40.2	38.2	37.1	35.5	31.8	43.6	39.7	44.6	41.4	32.7	46.8	35.2	43.5	31.8	40.9	40.2	38.1	37.9	37.8	CaO
3.2	1.7	2.5	2.8	1.9	2.5	3.1	2.9	1.6	2.1	1.7	2.1	2.7	1.8	1.4	2.1	1.6	1.7	0.8	2.6	1.8	1.3	1.2	MgO
0.31	0.75	0.75	0.27	0.26	0.75	0.32	0.29	0.27	0.75	0.75	0.36	0.75	0.28	0.75	0.31	0.75	0.32	0.75	0.75	0.75	0.75	0.31	Na_2O
1.2	0.7	0.6	1.1	1.4	0.7	1.7	0.7	1.1	0.3	0.5	0.3	0.4	0.4	0.4	0.6	0.7	1.7	0.4	1.1	0.7	0.6	0.5	K_2O
0.12	0.73	0.13	0.12	0.11	0.75	0.11	0.11	0.75	0.75	0.75	0.11	0.11	0.75	0.75	0.11	0.75	0.12	0.75	0.11	0.75	0.75	0.16	P_2O_5
0.75	0.11	0.75	0.75	0.73	0.11	0.75	0.11	0.75	0.75	0.75	0.16	0.75	0.11	0.18	0.12	0.73	0.11	0.75	0.12	0.19	0.11	0.75	SO_3
0.81	0.48	0.29	0.63	0.64	0.83	0.54	0.66	0.62	0.28	0.25	0.28	0.27	0.31	0.26	0.29	0.35	0.36	0.31	0.67	0.72	0.36	0.26	TiO_2
0.75	0.75	0.75	0.75	0.75	0.74	0.75	0.75	0.75	0.77	0.75	0.75	0.75	0.75	0.75	0.73	0.75	0.75	0.75	0.75	0.75	0.75	0.75	SrO
0.8	0.8	0.8	0.8	0.8	0.8	0.8	0.8	0.8	0.8	0.8	0.8	0.8	0.8	0.8	0.8	0.8	0.8	0.8	0.8	0.8	0.8	0.8	Cl

Table 4.4 Some results of oxide analysis by XRF in the Ramhormoz region in terms of percentage for training

R18	R17	R16	R15	R14	R13	R12	R11	R10	R9	R8	R7	R6	R5	R4	R3	R2	R1	No
21	20	21	21	19	20	21	22	22	8.8	10	20	23	23	21	22	19	15	SiO_2
4.1	4.6	5.2	3.6	3.1	2.7	2.6	2.8	2.6	1.8	2.1	4.4	5.8	5.3	4.9	4.9	5.3	3.6	Al_2O_3
3.3	1.7	3.1	2.9	3.7	3.2	4.1	3.8	3.2	1.3	1.8	3.1	4.1	3.9	3.5	3.3	3.9	2.8	Fe_2O_3
31.5	39.2	46.6	41.8	41.2	39.2	44.6	41.2	36.7	47.7	46.6	34.6	30.5	32.9	34.4	33.6	36.9	40.9	CaO
3.4	2.7	3.7	1.9	2.1	1.7	3.6	2.7	1.9	1.4	1.6	2.8	3.5	3.3	3.5	4	2.4	2.3	MgO
0.5	0.9	0.8	0.6	0.9	0.7	0.8	0.5	0.3	0.75	0.75	0.81	0.86	0.84	0.56	0.67	0.2	0.1	Na_2O
0.48	0.96	1	1.06	1.02	0.54	0.85	0.84	1.1	0.41	0.43	0.91	1.1	1	0.8	0.86	1.05	0.77	K_2O
0.16	0.13	0.1	0.11	0.1	0.75	0.76	0.18	0.13	0.1	0.1	0.75	0.1	0.75	0.75	0.1	0.11	0.13	P_2O_5
0.66	0.75	1.1	0.8	0.7	0.9	1.2	4.1	0.16	0.2	0.5	4.7	1.9	2.2	0.7	1.2	0.17	0.6	SO_3
0.72	0.54	0.66	0.42	0.38	0.35	0.48	0.37	0.32	0.75	0.75	0.47	0.44	0.44	0.5	0.5	0.52	0.34	TiO_2
0.2	0.3	0.4	0.2	0.1	0.4	0.2	0.3	0.1	0.1	0.21	0.31	0.21	0.27	0.1	0.1	0.1	0.42	SrO
0.9	0.9	0.8	0.6	0.8	0.5	0.8	0.8	0.7	0.8	0.8	0.8	0.5	0.5	0.9	1	0.8	0.2	Cl

Table 4.5 Some results of the geochemical analysis to determine sub-elements in the Dezful region, southwest of Iran

D23	D22	D21	D20	D19	D18	D17	D16	D15	D14	D13	D12	D11	D10	D9	D8	D7	D6	D5	D4	D3	D2	D1	No
15	17	30	10	16	17	21	30	10	17	16	20	10	17	38	29	18	30	15	16	21	10	17	**Ce**
8.4	12.1	9.78	11.7	9.24	8.6	12.1	9.71	11.5	20.1	9.21	12.2	8.2	11.3	20.2	16.7	11.6	16.7	9.23	9.76	12.2	8.7	11	**La**
7.25	2.86	4.74	10.4	4.16	7.91	3.37	3.2	4.2	4.73	10.3	2.85	7.22	4.17	10.3	7.22	4.2	7.91	3.37	3.2	4.73	2.85	4.16	**Sc**
14.5	8.95	6.33	7.25	13.5	12.5	8.15	8.23	7.16	8.96	13.1	6.32	7.22	8.15	14.4	12.5	8.23	13.1	7.22	7.16	8.96	6.32	8.15	**Y**
1452	941	2488	2589	1585	1458	1258	814	963	965	854	1316	788	1567	1225	1135	2628	1363	2478	3541	1958	1909	2309	**Ti**
8.51	7.69	12.54	7.15	8.66	<5.00	10.36	6.74	5.88	9.55	6.25	6.18	<5.00	6.73	<5.00	5.52	13.2	7.00	11.5	18.9	9.65	8.38	9.97	**Co**
141.8	135.5	65.7	85.3	72.9	70.6	150.3	142.6	96.5	75.2	60.2	59.2	46.5	73.5	55.9	58.4	120	65.6	114	160	90.6	80.6	100	**Zr**
42.2	30.2	65.1	74.3	36.1	25.1	45.6	92.3	86.4	52.3	29.8	38.7	31.0	40.2	28.2	32.7	63.8	42.6	59.3	98.1	48.3	44.0	53.8	**V**
2.45	1.78	1.36	2.85	3.52	1.21	3.54	2.36	1.58	2.58	1.52	2.07	1.35	2.46	1.88	1.68	4.30	2.15	3.87	4.68	3.52	3.00	4.16	**Th**
0.44	0.85	0.65	0.34	0.52	0.74	0.44	0.63	0.45	0.85	0.33	0.38	0.29	0.39	0.31	0.37	0.86	0.33	0.63	0.87	0.44	0.51	0.59	**Ta**
6.54	3.74	4.65	7.52	10.58	11.41	8.41	5.66	9.54	10.36	3.25	4.94	3.44	6.02	4.64	4.74	9.56	4.95	8.38	11.9	6.97	6.18	7.22	**Nb**

Table 4.6 Some results of the geochemical analysis to determine sub-elements in the Ramhormoz region, southwest of Iran

R18	R17	R16	R15	R14	R13	R12	R11	R10	R9	R8	R7	R6	R5	R4	R3	R2	R1	No
24.5	11.6	24.6	20.9	22.1	11.8	22.3	20.3	11.3	24.1	26.4	27	25.4	24.1	22.5	20.4	11.5	16.6	Ce
15.4	7.5	13.5	14.3	10.2	15.2	13.4	7.2	10.4	14.6	15.1	15.4	14.6	15.4	14	13.4	7.3	10.5	La
6.75	2.44	4.45	2.95	5.69	6.91	2.47	4.46	2.95	4.96	6.75	5.81	5.53	5.69	6.91	4.46	2.47	2.98	Sc
12	5.8	11	12	7.4	11	5.8	11	7.3	10	12	11	11	10	11	10	5.8	7.3	Y
1256	1478	856	1632	1584	2185	753	826	662	1434	1984	2026	2029	2108	2132	1705	1161	1597	Ti
<5.00	6.71	9.69	7.12	8.63	10.25	6.78	<5.00	<5.00	6.66	10.6	8.02	8.87	9.98	10.4	6.96	<5.00	7.56	Co
92.4	65.7	85.2	74.3	58.3	82.7	47.2	51.4	43.1	83.4	116	87.1	85.9	92.6	101	75.1	51.3	68.6	Zr
34.1	28.1	64.5	32.5	42.8	65.4	29.5	28.3	29.1	43.0	66.6	45.7	47.1	49.2	55.3	39.0	27.6	37.4	V
1.69	1.78	2.52	3.54	3.61	1.68	1.54	1.73	1.35	2.68	3.38	2.78	3.01	3.52	3.60	3.03	1.87	2.78	Th
0.41	0.51	0.31	0.28	0.32	0.42	0.25	0.27	0.20	0.40	0.50	0.43	0.39	0.36	0.47	0.48	0.30	0.57	Ta
3.32	4.25	2.85	5.36	7.25	3.65	2.96	3.99	2.79	5.21	6.68	5.87	6.22	6.23	7.69	6.02	4.62	5.66	Nb

The average elements of Er, Eu, Ho, Lu, Tb, Tm and Yb in the samples of the Dezful study area are less than one ppm (Table 4.7).

Also, the most abundant rare earth elements in the Ramhormoz region include Ce with an average of 21.12 ppm, followed by La with an average of 12.4 ppm, Y with an average of 9.11 ppm, Nd with 9.3 ppm and Sc with an average of 4.7. Ppm is Pr with 2.76 ppm, Sm with an average of 2.06 ppm, Gd with an average of 1.8 ppm and Dg with an average of 1.56 ppm. The average elements of Er, Eu, Ho, Lu, Tb, Tm and Yb are similar to the Dezful region in the samples of the Ramhormoz study area are less than one ppm (Table 4.8).

These elements are specified in percentages 4–16. As can be seen, the most abundant rare earth elements in both Ramhormoz and Dezful regions are Ce, La, Nd, Sc and Y (Fig. 4.5).

4.2.4 Discussion

The results of elementary studies of Andimeshk and Ramhormoz regions show a lot of information in terms of their different sedimentary, environmental and tectonic properties that can be analyzed with different geochemical methods. In this section, the aim is to discuss the different properties of these sediments based on the results of the geochemical data of these sediments.

One of the most efficient methods of geochemical studies is the elemental comparison between two different parts in terms of the frequency of oxides present (Abu et al. 2020). In fact, the abundance of different elements in each environment indicates the conditions of formation and the type of environment that constitutes it (Oreshkina et al. 2020). Since in both Ramhormoz and Dezful regions Cao and SiO_2 are the most abundant elements, in this section, it has been tried to study the changes of these two oxides first at the harvest points and then to make a comparison between these two regions. According to the studies, the changes of SiO_2 in the Ramhormoz region have a more uniform trend than the changes of this oxide in the Dezful region (Fig. 4.6a). This may indicate that the sediment maturity in the samples of the Ramhormoz region is uniform compared to the samples of the Dezful region (Sahraeyan and Bahrami 2016).

The changes in calcium oxide in both regions have a trend of changes almost similar to each other (Fig. 4.6b). Comparing the tables in Fig. 4.6a and b, we find that in addition to the abundance of calcium oxide in each of the two regions, the trend of changes in silicon oxide is opposite to the trend of changes in calcium oxide in these regions (Abu et al. 2020).

4.2.4.1 Study of Main Elements Based on the Aluminum Scale

Aluminum is usually a stable element in nature and often increases in humid climates and decreases in dry climates (Heidari and Raheb 2020). In fact, aluminum oxide

Table 4.7 Frequency of rare earth elements in the Dezful region in ppm

D23	D22	D21	D20	D19	D18	D17	D16	D15	D14	D13	D12	D11	D10	D9	D8	D7	D6	D5	D4	D3	D2	D1	No
15	17	30	10	16	17	21	30	10	17	16	20	10	17	38	29	18	30	15	16	21	10	17	Ce
2.36	2.19	1.53	2.18	1.38	1.54	2.18	2.16	1.37	1.52	2.31	1.35	2.15	1.36	2.58	2.16	1.39	2.36	1.2	1.27	1.56	1.07	1.38	Dy
0.9	0.8	1.5	0.7	1.3	0.8	1	0.7	1	0.7	1.3	0.9	0.7	0.8	1.5	1.3	0.8	1.3	0.7	0.7	0.9	0.7	0.8	Er
0.35	0.41	0.67	0.34	0.32	0.88	0.43	0.87	0.34	0.41	0.66	0.31	0.35	0.42	0.88	0.59	0.39	0.67	0.35	0.35	0.47	0.3	0.41	Eu
1.56	1.67	1.33	1.54	2.62	3.15	1.34	1.66	1.68	1.67	1.32	3.14	2.63	1.65	3.15	2.61	1.68	2.65	1.34	1.54	1.9	1.37	1.66	Gd
0.4	0.3	0.4	0.4	0.6	0.2	0.4	0.3	0.5	0.3	0.6	0.5	0.3	0.3	0.6	0.4	0.3	0.5	0.3	0.3	0.4	0.2	0.3	Ho
8.4	12.1	9.78	11.7	9.24	8.6	12.1	9.71	11.5	20.1	9.21	12.2	8.2	11.3	20.2	16.7	11.6	16.7	9.23	9.76	12.2	8.7	11	La
0..13	0.8	0.9	0.13	0.21	0.8	0.14	0.9	0.12	0.13	0.12	0.2	0.14	0.12	0.23	0.21	0.12	0.22	0.1	0.12	0.14	0.09	0.13	Lu
7.82	17.2	8.76	9.96	6.98	7.84	8.74	8.84	14.2	6.62	6.93	7.85	9.97	8.85	17.1	14.2	8.76	14.2	6.61	7.85	9.96	6.93	8.83	Nd
2.48	1.88	3.94	2.78	2.19	2.6	3.88	2.46	1.9	3.96	2.72	2.18	1.82	2.4	4.3	3.88	2.46	3.96	1.9	2.18	2.72	1.81	2.5	Pr
7.25	2.86	4.74	10.4	4.16	7.91	3.37	3.2	4.2	4.73	10.3	2.85	7.22	4.17	10.3	7.22	4.2	7.91	3.37	3.2	4.73	2.85	4.16	Sc
3.4	1.38	2.99	1.99	1.85	3.6	2.93	1.56	1.76	1.39	1.42	3.03	1.98	1.82	3.6	2.92	1.76	3.03	1.42	1.56	1.98	1.39	1.81	Sm
0.2	0.44	0.21	0.26	0.45	0.25	0.28	0.8	0.21	0.25	0.28	0.8	0.44	0.25	0.5	0.4	0.25	0.44	0.21	0.2	0.28	0.2	0.26	Tb
0.2	0.22	0.13	0.11	0.25	0.09	0.22	0.04	0.07	0.14	0.21	0.1	0.09	0.13	0.25	0.22	0.13	0.21	0.1	0.09	0.13	0.09	0.13	Tm
14.5	8.95	6.33	7.25	13.5	12.5	8.15	8.23	7.16	8.96	13.1	6.32	7.22	8.15	14.4	12.5	8.23	13.1	7.22	7.16	8.96	6.32	8.15	Y
0.75	1.45	0.78	1.45	1.26	1.48	0.58	1.24	0.66	0.86	0.72	0.64	1.41	0.77	1.48	1.24	0.72	1.41	0.66	0.58	0.86	0.64	0.77	Yb

Table 4.8 Frequency of rare earth elements in the Ramhormoz region in ppm

R18	R17	R16	R15	R14	R13	R12	R11	R10	R9	R8	R7	R6	R5	R4	R3	R2	R1	No
0.2	0.4	0.4	0.3	0.4	0.4	0.3	0.3	0.3	0.3	0.3	0.2	0.3	0.2	0.4	0.4	0.4	0.4	**Cd**
24.5	11.6	24.6	20.9	22.1	11.8	22.3	20.3	11.3	24.1	26.4	27	25.4	24.1	22.5	20.4	11.5	16.6	**Ce**
1.9	1.2	1.4	1.7	1.1	1.2	2	1.6	1.2	1.6	2	2	1.8	1.8	1.9	1.7	1	1.2	**Dy**
0.92	1.02	1.06	0.66	1.08	1.12	0.91	1.02	0.62	0.92	1.05	1.11	1.03	1.01	1.09	0.96	0.51	0.68	**Er**
0.3	0.5	0.2	0.4	0.7	0.4	0.5	0.3	0.4	0.5	0.6	0.6	0.5	0.5	0.5	0.5	0.3	0.4	**Eu**
1.9	1.1	2.3	1.5	2.2	2.1	1.1	1.9	1.4	2	2.3	2.3	2	2.3	2.1	1.9	1.1	1.5	**Gd**
0.12	0.41	0.21	0.24	0.46	0.29	0.32	0.18	0.23	0.33	0.43	0.41	0.39	0.33	0.33	0.29	0.18	0.22	**Ho**
15.4	7.5	13.5	14.3	10.2	15.2	13.4	7.2	10.4	14.6	15.1	15.4	14.6	15.4	14	13.4	7.3	10.5	**La**
0.17	0.07	0.16	0.12	0.05	0.15	0.18	0.09	0.11	0.16	0.19	0.17	0.15	0.16	0.17	0.14	0.09	0.11	**Lu**
9.54	5.2	12.7	11.8	7.85	10.9	9.55	5.5	7.88	10.9	12.6	11.7	11.3	11.8	10.1	9.54	5.6	7.89	**Nd**
3.5	2.9	2.4	3.2	3.2	2.8	3.1	1.8	2.4	3.1	3.3	3.3	3.2	3.1	3	2.8	1.7	2.4	**Pr**
6.75	2.44	4.45	2.95	5.69	6.91	2.47	4.46	2.95	4.96	6.75	5.81	5.53	5.69	6.91	4.46	2.47	2.98	**Sc**
1.5	2.25	1.62	2.36	1.4	2.33	1.66	2.21	1.3	2.34	2.54	2.54	2.35	2.34	2.21	2.06	1.2	1.63	**Sm**
0.55	0.42	0.37	0.24	0.39	0.36	0.52	0.41	0.26	0.48	0.47	0.36	0.39	0.43	0.5	0.4	0.2	0.27	**Ta**
0.18	0.32	0.33	0.21	0.16	0.32	0.33	17	0.2	0.28	0.35	0.33	0.33	0.33	0.32	0.28	0.16	0.2	**Tb**
0.2	0.1	0.1	0.1	0.2	0.2	0.1	0.1	0.1	0.2	0.2	0.2	0.2	0.2	0.2	0.2	0.1	0.1	**Tm**
12	5.8	11	12	7.4	11	5.8	11	7.3	10	12	11	11	10	11	10	5.8	7.3	**Y**
0.51	0.96	1.05	0.63	1.03	0.53	0.98	0.51	0.65	0.9	1.05	1.09	0.98	0.9	1	0.86	0.51	0.64	**Yb**

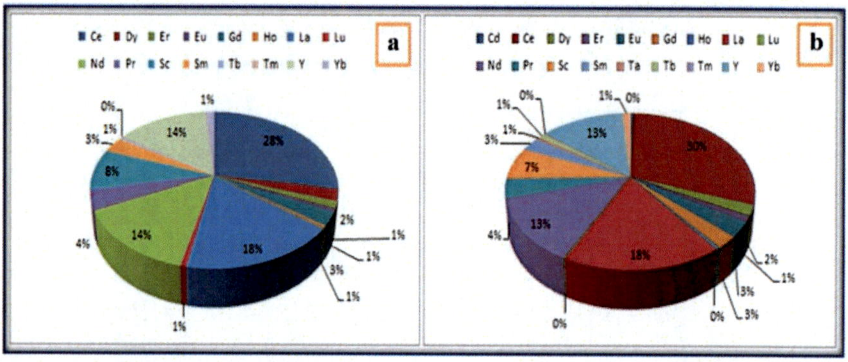

Fig. 4.5 Average percentage of the abundance of rare earth elements in the studied areas (based on percentage). **a** The average percentage of rare earth elements in the Dezful region. **b** The average percentage of rare earth elements in the Ramhormoz region

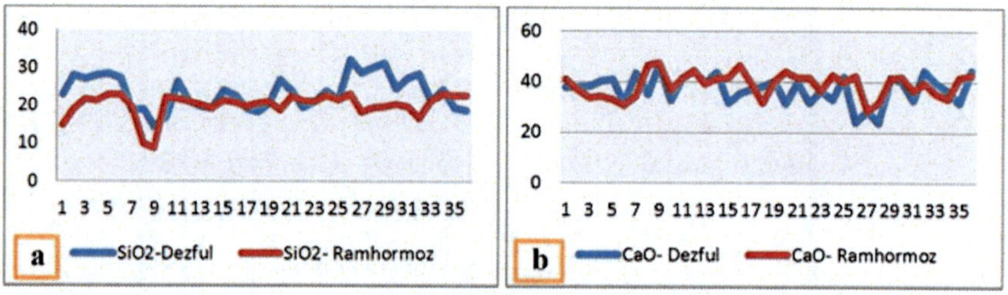

Fig. 4.6 Elemental comparison of Ramhormoz and Dezful regions based on the percentage (sample number (row) versus percentage (column)). **a** Graph of frequency changes of silicon oxide in Dezful and Ramhormoz regions in samples no. 1 to 35. **b** Graph of frequency changes of calcium oxide percentage in Dezful and Ramhormoz regions in samples 1 to 35

is considered as a constant factor during diagenesis, weathering and metamorphism (Smirnov et al. 2019). Therefore, this factor is used as a scale in the study of the main elements in detrital sediments (Sahraeyan and Bahrami 2016). However, the elements CaO, Na_2O and K_2O in sandstones are known as the most mobile phases (Sharma et al. 2020).

The positive correlation between SiO_2 and AL_2O_3 can be attributed to the presence of these elements in quartz and other minerals. In fact, the SiO_2/AL_2O_3 ratio is used as an index to determine sediment maturity (Smirnov et al. 2019). This handling increases during weathering, transport and recycling, resulting in an increase in quartz relative to unstable components such as feldspar and rock fragments. SiO_2/AL_2O_3 ratio higher than 5 to 6 in sedimentary rocks indicates high sediment maturity (Roser and Korsch 1986). This positive association can indicate the presence of feldspars, mica and clay minerals (Sorokin et al. 2017).

In the studied samples, the average percentage of aluminum oxide in the sediments of the Dezful region is about 5.27% and in the sediments of the Ramhormoz region is about 4.03%. In these areas, the oxide shows a positive trend with Fe_2O_3, K_2O, MgO, TiO_2 and to some extent with SiO_2 and a negative correlation with CaO oxide (Fig. 4.7).

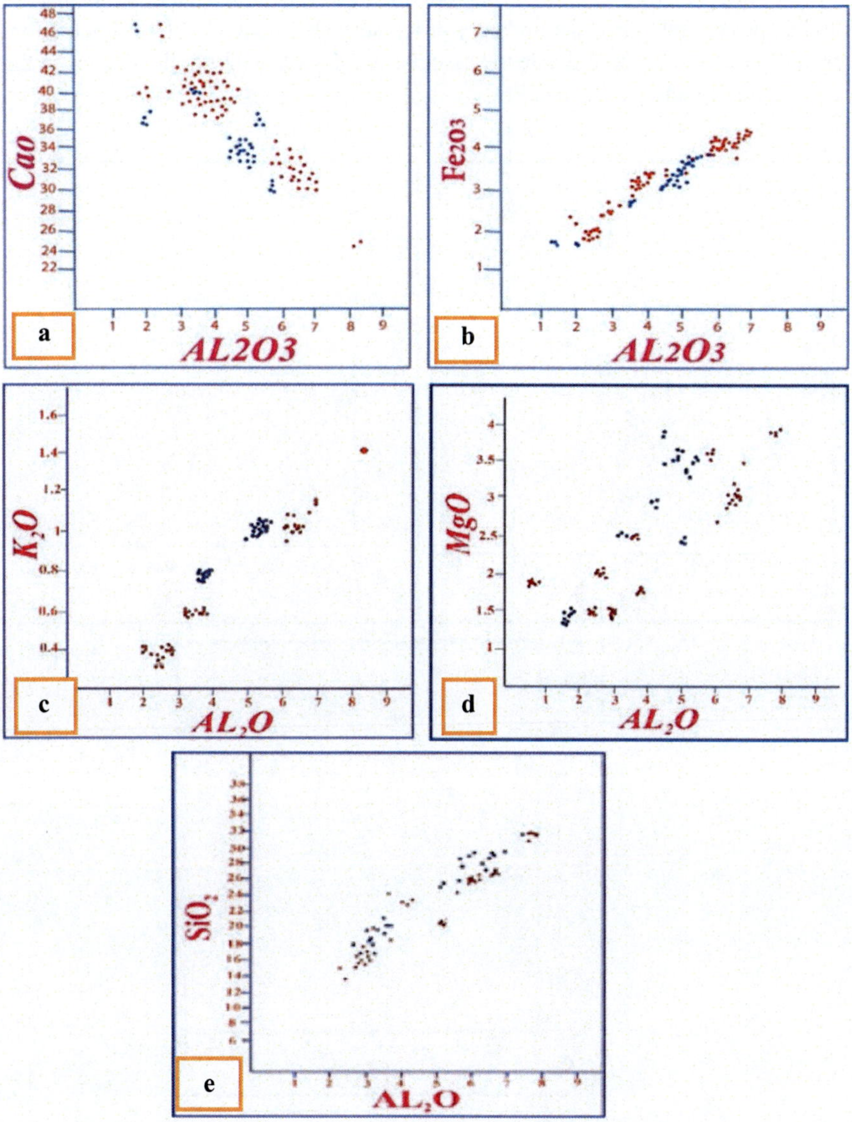

Fig. 4.7 Investigation of changes of main oxides with aluminum oxide (blue points of the Ramhormoz region and red points of the Dezful region). **a** Negative relationship between AL_2O_3 and CaO. **b** Positive relationship between AL_2O_3 and Fe_2O_3. **c** Positive relationship between AL_2O_3 and K_2O. **d** positive association between AL_2O_3 and MgO. **e** Positive association between AL_2O_3 and SiO_2

Examination of the changes of TiO_2 with AL_2O_3 shows that these two oxides have a trend of positive changes relative to each other (Fig. 4.8a). This oxide has an average of 0.52% in the Dezful region and 0.53% in the Ramhormoz region. The element titanium is more concentrated in filosilicates and is a good indicator for the interpretation of source rock compared to other elements due to inactivity and displacement during sedimentary processes (Sharma et al. 2020). Interaction of AL_2O_3 with TiO_2 may indicate the association of TiO_2 with phyllosilicates in these areas (Heidari and Raheb 2020).

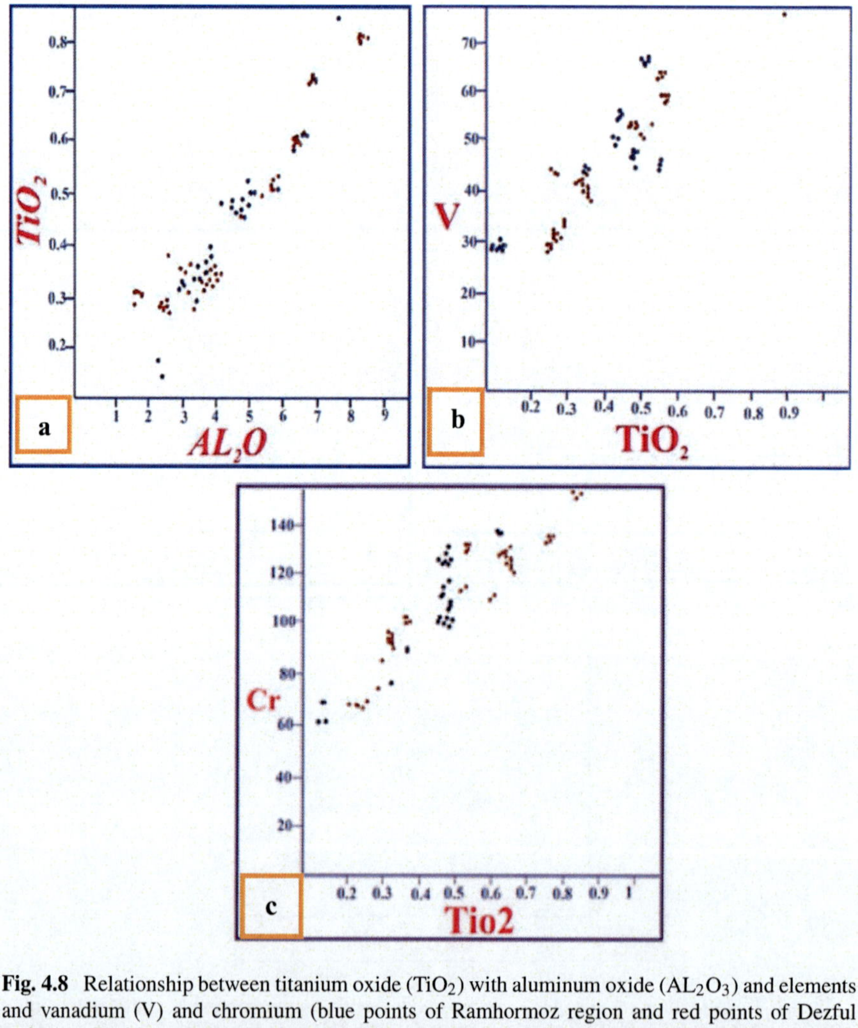

Fig. 4.8 Relationship between titanium oxide (TiO_2) with aluminum oxide (AL_2O_3) and elements and vanadium (V) and chromium (blue points of Ramhormoz region and red points of Dezful region). **a** Positive relationship between TiO_2 and AL_2O_3. **b** Positive relationship between TiO_2 and V. **c** Positive relationship between TiO_2 and Cr

In addition, the study of titanium oxide changes with v and cr elements also indicates a positive relationship between this oxide and the mentioned elements (Fig. 4.8b, c). In addition, the correlation of cr and v with TiO_2 oxide indicates the presence of heavy minerals in these areas because cr and v are associated with iron and titanium in heavy minerals (Al-Hashim and Corcoran 2020). The presence of heavy minerals with petrographic data has been confirmed in these studies.

Sub-elements such as Ti, Nb, Ta, Cs, Ce, Ni, V, Y, Zr, Sc, Th, La and Co are generally their primary properties due to their immobility during weathering, diagenesis and metamorphism which are preserved in sedimentary rocks so these sub-elements can be used to determine the origin and location of tectonics (Sharma et al. 2020). A comparison of trace elements with aluminum oxide AL_2O_3 shows that the elements such as scandium (Sc), vanadium (v), thorium (Th), cobalt (Co), niobium (Nb), chromium (Cr), lanthanum (La) with aluminum oxide have the same correlation process (Fig. 4.9).

The existence of a positive correlation between these elements, especially the positive correlation between AL_2O_3 and Cr, is a reason for the presence of heavy minerals in these areas, the existence of these heavy elements has been confirmed in petrographic studies (Heidari and Raheb 2020)

4.2.4.2 Geochemical Classification of Sediments

In addition to petrographic methods, researchers use geochemical methods, especially the composition of the main elements in the classification and separation of mature and immature sediments (Al-Hashim and Corcoran 2020).

Geochemical Classification of Sandstones Based on Herron Index (1988)

The log diagram (Fe_2O_3/K_2O) versus log (SiO_2/Al_2O_3) is used to classify sandstones (Herron 1988). The location of data related to the studied sections (from sandstone samples taken from Aghajari and Bakhtiari Formations in the mentioned areas) on this diagram shows that in the Dezful region, most of the samples are in the range of light arnite to sand. Iron ores are found, as evidenced by the petrographic data previously discussed (Fig. 4.10a). The studied samples in the Ramhormoz area are mainly in the range of Arnite limestone to iron-bearing sands and iron-bearing shales (Fig. 4.10a).

Geochemical Classification of Sandstones Based on the Index of Petty John et al. (1987)

The log diagram (Na_2O/K_2O) versus log (SiO_2/Al_2O_3), which is based on chemical maturity indices, is used to classify detrital sandstones (Pettijohn et al. 1987). Drawing the results of the analysis of the main epoch in Fig. 4.10b shows that most

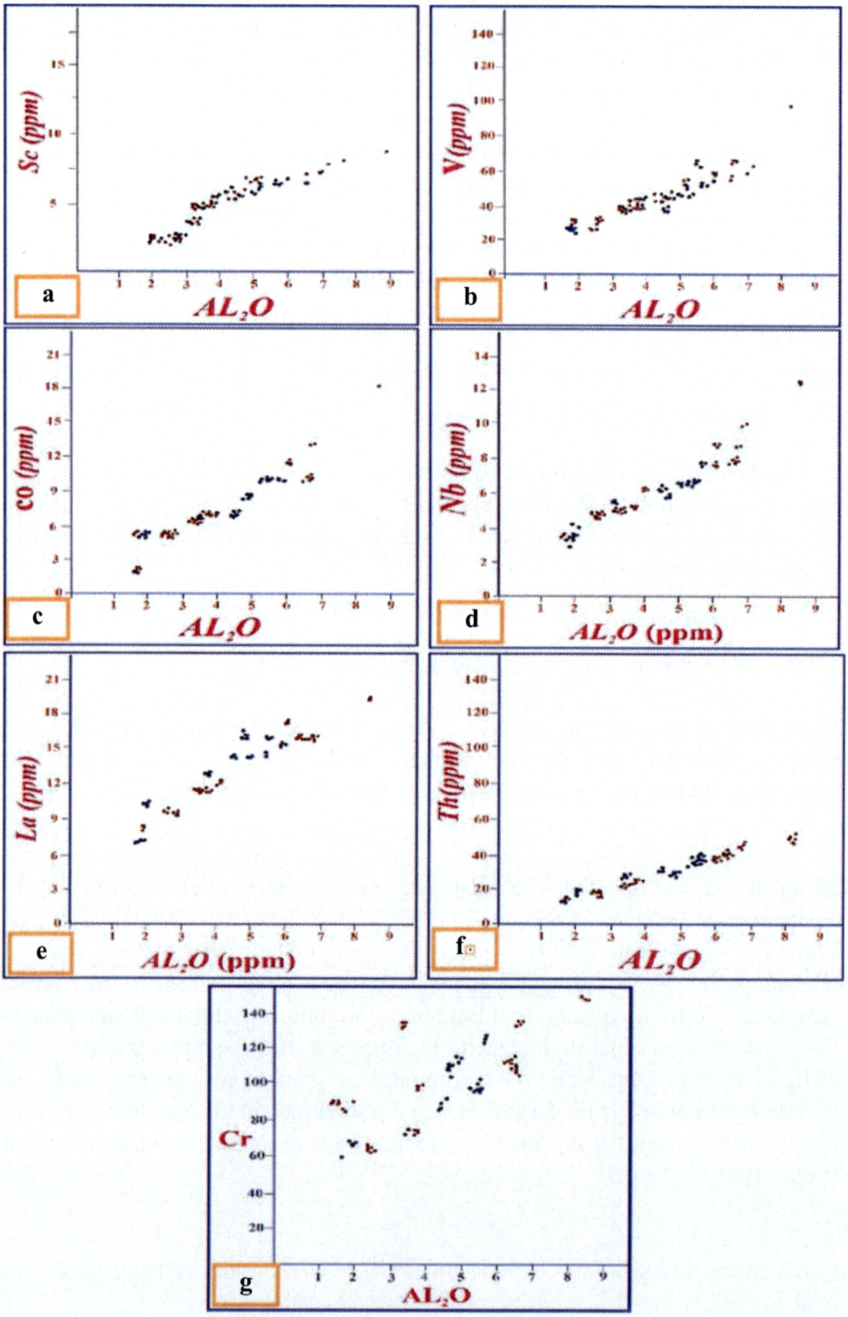

Fig. 4.9 Relationship between aluminum oxide and by-elements in the studied samples

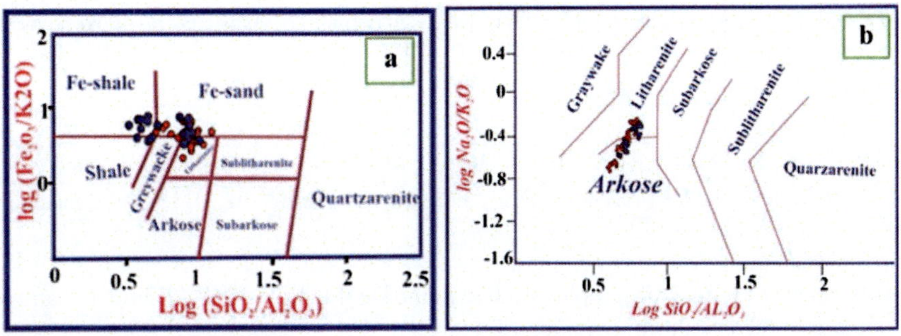

Fig. 4.10 Geochemical classification of sediments in Dezful (red points) and Ramhormoz (blue points). **a** Geography of the studied samples on the diagram presented by Herron, 1988. **b** Geochemical classification of the studied samples on the Pettijohn, 1987 diagram

of the sandstone samples in the Dezful region have a light arnitecomposition and a small amount of archetype and in the Ramhormoz region mainly the samples have an arnite light composition. These results are also confirmed by petrographic data (Fig. 4.10b).

4.2.4.3 Initial Weathering Studies of Sediments

Geochemistry of sedimentary rocks is a complex function of the nature of sedimentary rocks, intensity and weathering period, re-sedimentation cycle, diagenesis and sorting (Oreshkina et al. 2020). Geochemical analysis of the main elements of detrital rocks is a suitable tool for determining the tectonic position in sandstones that have not been strongly affected by diagenesis and metamorphism or other alteration and weathering processes (Smirnov et al. 2019). Therefore, it is necessary to study the role of weathering and determine diagenesis to study and evaluate the types of sediment origins (Oreshkina et al. 2020).

The weathering history of rocks is often estimated by calculating the ratio of the mobile oxides CaO, K_2O and Na_2O to the non-mobile oxide AL_2O_3. The most widely used index in this regard is the Chemical Index of Alteration (CIA) as proposed by Nesbitt and Young (1982). This index is obtained by the following formula, and oxides are expressed in molar ratio.

$$CIA = \left[AL_2O_3/\left(AL_2O_3 + CaO^* + Na_2O + K_2O\right)\right] \times 100$$

Calcium is present in silicate components, and this amount may be low, medium or high. An increase in the CIA value from the bottom up is associated with the degree of chemical alteration (Smirnov et al. 2019). Medium and high are associated with the transfer of mobile cations (such as Ca^{2+}, Na^{+}, K^{+}) and the persistence of constituents (Al^{3+}, Ti^{+}) with less mobility (Oreshkina et al. 2020). The average of

Table 4.9 Average CIA index for Dezful (A) and Ramhormoz (B) study areas. (The exact position of the samples is presented in the sedimentology chapter)

D45	D40	D35	D30	D25	D20	D15	D10	D5	D1	NO
16.194	24.679	15.019	8.469	17.326	5.877	5.109	8.089	4.256	7.817	Dezful
R24	R21	R18	R15	R12	R9	R6	R3	R2	R1	No
14.126	13.155	12.343	10.789	15.045	13.322	12.059	12.189	12.309	7.86	Ramhormoz

this index is 11.2 for 10 samples studied in the Dezful region and 12.3 for 10 samples studied in the Ramhormoz region (Table 4.9).

Sometimes the use of the CIA index in samples that show large changes in CaO does not provide interesting results (Smirnov et al. 2019). This can be seen in the samples of both regions so that the high amount of CaO (due to the presence of particulate matter) has caused a decrease in the average CIA, which should be considered in the final results and interpretation of weathering.

Therefore, in this regard, Cullers and Podkovyrov (2002) presented another weathering index for sandstones with variable CaO content. This index is defined as follows:

$$CIW' = [AL_2O_3/(AL_2O_3 + Na_2O)] \times 100$$

In this formula, oxides should be considered as molecular ratios. The range of variation of this index is from 50 for non-aerated sediments and 100 for highly aerated sediments.

Thus, this index has been obtained for the samples of the Dezful region with an average of 71 and in the Ramhormoz region with an average of 77. This number indicates the long-distance of sediment transport from the origin to the sedimentation area and the humid climate of the source area (Zagros highlands) for these samples (Oreshkina et al. 2020).

Considering the mean and range of changes of this index for the studied samples, we suggest a moderate amount of weathering for these samples. The moderate weathering obtained for these samples may indicate intense re-sedimentation cycle activity in humid to semi-humid climates or the presence of re-cycling in semi-arid to arid climates.

Therefore, the obtained CIW geochemical index of weathering may be the result of recirculation in semi-arid to arid climates (Smirnov et al. 2019). Of course, accurate access to climatic conditions in relation to these deposits requires more detailed studies, such as the study of clay minerals.

Examination of the results with the climatic results of the source regions regarding sediments in the Dezful region (in Lorestan province) and sediments in the Ramhormoz region (in Chaharmahal and Bakhtiari province) shows that the data obtained in these studies are in line with the current climatic conditions. It is consistent in these areas. In fact, Lorestan province is divided into three climatic regions in terms of climatic conditions, which include the following (LashniZ et al. 2011):

arid and cold mountainous areas (relatively high areas—more than 1400 m). Areas include the cities of Boroujerd, Aleshtar, Noorabad, Aligudarz and Azna.

- Central semi-arid and temperate region: This region has a lower altitude and includes the cities of Dasht Mountain, Khorramabad.
- Hot and dry to hot humid southern region: The territory of this climate includes the lowest areas of Lorestan. High average annual temperature, severe and permanent evaporation and little rainfall are the main features of this climatic zone. And includes the Old Office, Papi and adjacent areas.

As can be seen, the data obtained from geochemical studies in the Dezful study area, which indicate the semi-arid to arid climatic conditions for this region, are consistent with the current climatic conditions of the regions of origin of these sediments (mainly Lorestan province).

In addition, regarding the climatic conditions of the upstream areas of Ramhormoz (Chaharmahal and Bakhtiari province), 4 climates have been introduced as follows (Arvin et al. 2016):

- Moderate and very cold humid climates are located in the areas adjacent to the heights and occupy a small area of the entire province.
- Cold temperate semi-humid climate: covers a large area of the province. The central regions of the province and many plains are located in this climate, which includes Shahrekord city including central parts, Shalmzar, Laran, Ben and part of Saman part and in Borujen city central part and Farsan city and Nagan city in Ardal city.
- Cold temperate humid climate: This climate is almost below the third type of climate (cold temperate semi-humid) with the difference that the amount of rainfall is more in this climate. This climate includes the areas of Imam Qais in the city of Borujen, the dinars of Ardal and Sudjan, and Morghmalek in the city of Shahrekord.
- Semi-humid to semi-arid climate: includes Lordegan areas and Bazaft section in Koohrang city and Zayandehrud margin in Saman section.

As can be seen, the mentioned data confirm that the results of geochemical studies are consistent with the current conditions of the source of Ala Ramhormoz rivers, which are mainly from Chaharmahal and Bakhtiari provinces (Lordkan and Koohrang counties).

In the A-CN-K diagram, the weathering path can be obtained from both weathering profile method and the thermodynamic estimation method (Nesbitt and Young 1982). In this diagram, the initial stages of weathering will have a parallel trend on the A-CN side because during the early stages of weathering, sodium and potassium ions are harvested by degradation of feldspars (plagioclase), continuing the weathering process of potassium feldspar (K) feldspars, its potassium ions are also reduced and the weathering process shifts to the AL_2O_3 compound (Paikaray et al. 2008). That is, in advanced weathering, a sharp decrease in K_2O occurs as the composition of these elements moves toward vertex A (Nesbitt and Young 1982). The weathering trend of the studied samples in the Dezful region (Fig. 4.11) is shown.

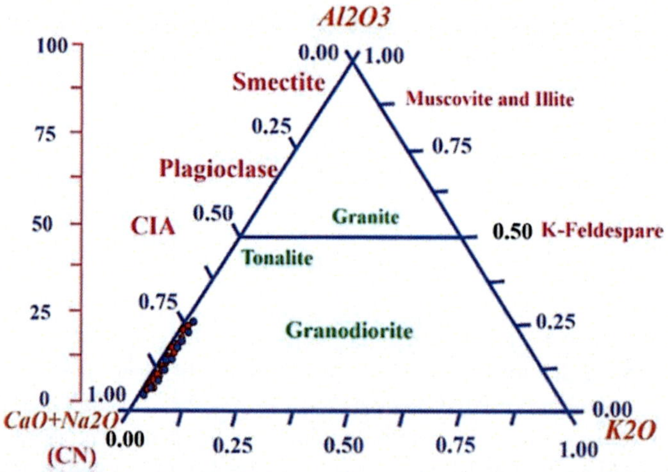

Fig. 4.11 Drawing samples of the studied areas in the combined space of the A-CN-K diagram and its extent to the connecting line of feldspars (Paikaray et al. 2008). Red points for the Dezful region and blue points for the Ramhormoz region

The drawing of samples in the composite space of Paikaray et al. (2008) shows that the data are close to or parallel to the A-CN line. Also, according to the studies of Hessler and Lowe (2006), the lack of drawing of samples near the AL_2O_3–K_2O connecting line indicates that there are no very severe weathering conditions in the mother rock area. This is true for samples of the target areas with moderate weathering, which are far from the Al_2O_3 pole and the Al_2O_3–K_2O junction and close to the feldspar junction (Fig. 4.11).

The drawing of the samples in this diagram shows that the samples in both scattering regions and a wide range in relation to the feldspar connecting line do not show in the diagram that such a trend indicates severe chemical weathering or the presence of a recirculation cycle. It is in detrital sediments (Abu et al. 2020). Thus, the homogeneous composition of the samples of these two ranges indicates weathering with a uniform range, where the rate of erosion and weathering is in equilibrium, and according to Nesbitt and Young (1982) such evidence indicates weathering mother rock in the climate. It is hot and humid (Oreshkina et al. 2020).

Such climatic conditions are associated with the transfer of mobile cations (such as Ca^{2+}, Na^{+}) and the persistence of constituents (Al^{3+}, Ti^{+}) with less excitation (Abu et al. 2020) which reduces the abundance of clay feldspars and minerals, the amount of which is in control of the amount of elements K_2O and Na_2O, Ca (mobile phases).

4.2.4.4 Climate Studies

By plotting the amount of SiO_2 against $AL_2O_3 + K_2O + Na_2O$ (Suttner and Dutta 1986), it is possible to understand the ancient climatic conditions in the source area

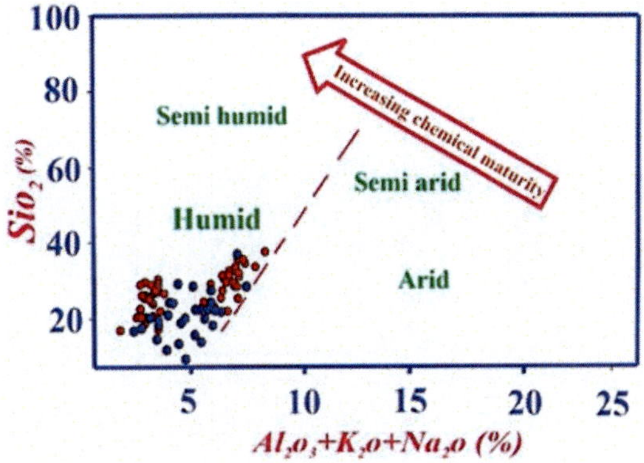

Fig. 4.12 Diagram of determining the ancient weather conditions in Dezful and Ramhormoz regions based on the diagram (Suttner and Dutta 1986)

(the source of the river). According to this diagram (Fig. 4.12), the ancient climate in the Mansha region was humid at the time of deposition in the Dezful region and the Ramhormoz region. These results are completely consistent with the results of the chart (Paikaray et al. 2008). As previously stated, the humid climate conditions expressed for both study areas with the current climate conditions of Lorestan and Chaharmahal and Bakhtiari provinces are completely consistent.

4.2.4.5 Determining the Tectonic Position of Sediments

The chemical composition of detrital sediments is affected by factors such as sedimentation processes, diagenesis and sediment composition, so the main elements of these sediments can be used to determine their tectonic position (Afarin et al. 2010). By changing the values of the main elements, detrital sediments of oceanic arc islands, continental arc islands, active continental margin and inactive margin can be separated from each other (Bhatia 1986)

In this section, before examining the geochemical data to determine the tectonic position of the study areas, first, by petrographic studies based on Dickinson (1985) diagram, the tectonic position of the study areas is determined and then it is matched with geochemical data.

In general, the composition of sand sediments depends on the characteristics of the origin, transport distance and changes after sedimentation, and the main relationship between the origin and sedimentation basin is controlled by tectonic processes, which place the main components of sandstones (quartz, feldspar). And particulate matter) in the source diagrams, the relationship between tectonic position and sedimentation can be determined (Jin et al. 2006). In this study, the diagrams of Qt, F, L and Qm, F, L Dickinson (1985) (Figs. 4.13 and 4.14) have been used.

After placing the percentage of the main components of the sandstone, the origin of the Undissected Arc for all the samples of the Ramhormoz and Dezful region has been determined in all three Dickinson (1985) diagrams. The location of the mentioned

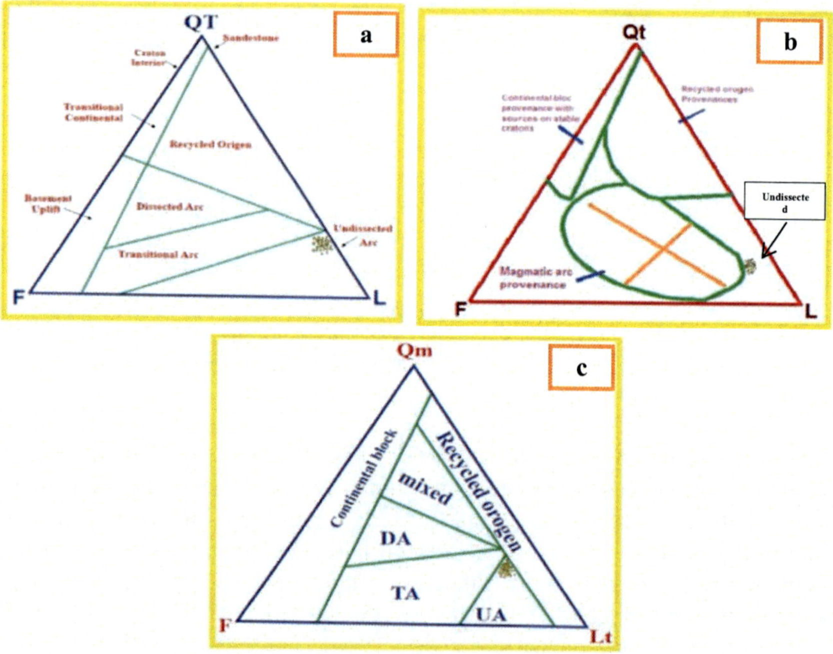

Fig. 4.13 Diagrams (**a**, **b**) Qt, F, L and Qm, F, Dickinson (1985) to determine the tectonic position of the studied sand samples in the Dezful region, which indicate the origin of Undissected Arc for these samples

samples in the mentioned range indicates the high percentage of carbonate particles in these sediments.

Determining the tectonic position of sediments in the Dezful and Ramhormoz regions according to the Bahatia diagram (Bhatia 1986) shows that the samples of Ramhormoz and Dezful regions are more inclined to the continental arc islands and the active continental margin (Fig. 4.15). The reason for the displacement of the samples relative to the specified ranges of Bhatia (Bhatia 1986) can be the presence of Fe_2O_3, MgO oxides in the network of carbonate particles in the studied sand samples (Heidari and Raheb 2020).

In general, the study of Fig. 4.15a–c, which is used to study the tectonic position of sedimentary samples, shows that the sedimentary samples of the Dezful region are mainly in the Active continental margin (probably From the Sanandaj zone of Sirjan) and to a lesser extent in the Continental island Arc area (probably from the Urmia Dokhtar zone) but the samples of Ramhormoz area are mainly in the Active continental margin area. As can be seen, the mentioned data are consistent with the data obtained from petrographic studies.

Due to their proximity to the source, the mentioned sediments can be formed from the re-cycle of sediments after orogeny and have entered the basin (Afarin et al. 2010). Sub-elements such as La, Th, Sc, Zn, Ti, Y and Cr are less affected by immobility

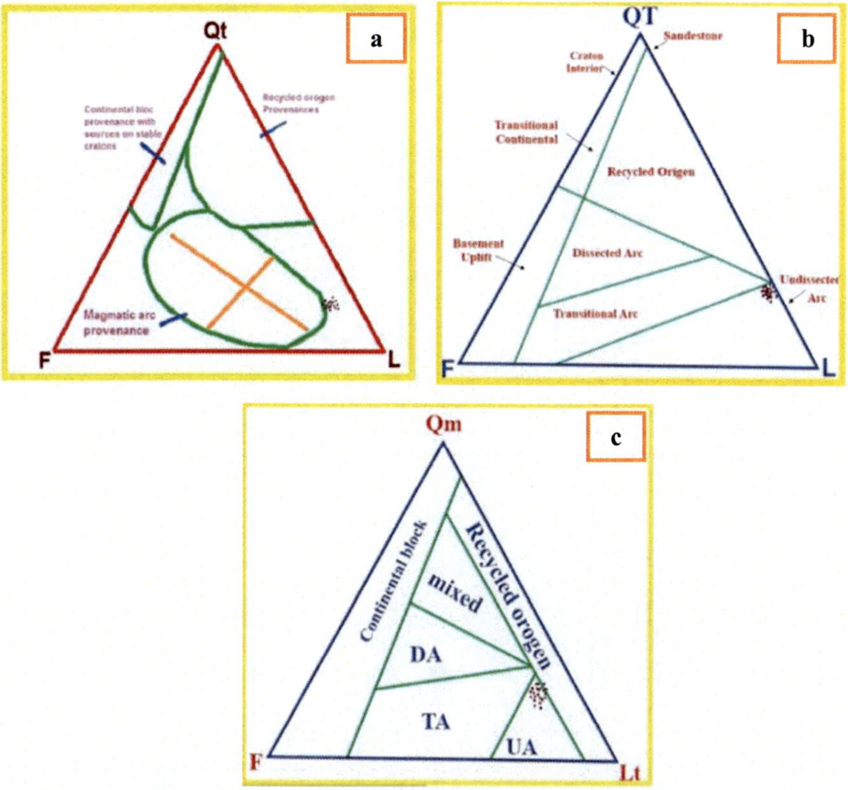

Fig. 4.14 Charts Qt, F, L (**a**, **b**) and Qm, F, Dickinson (1985) to determine the tectonic position of the studied sand samples in the Ramhormoz area, which indicates the origin of Undissected Arc for these samples

under aerobic conditions, diagenesis and moderate metamorphism (Fig. 4.15c) and therefore in the sample. Sedimentary rocks are preserved and are good indicators for understanding the long-standing tectonic position of sedimentary rocks (Grew 2009).

In addition to the diagrams provided to determine the tectonic position of the detrital specimens of the Dezful and Ramhormoz regions, the diagram (Schandl and Gorton (2002) was used based on sub-elements. As can be seen from this diagram, all specimens in the Dezful and Ramhormoz regions are located within the active continental margin (ACM) (Fig. 4.16a, b).

The study of Roser and Korsch (1986) diagram to determine the geological location of the samples in the study areas of Dezful and Ramhormoz also confirms the geological location of the Active Continental Margin for these samples (Fig. 4.16).

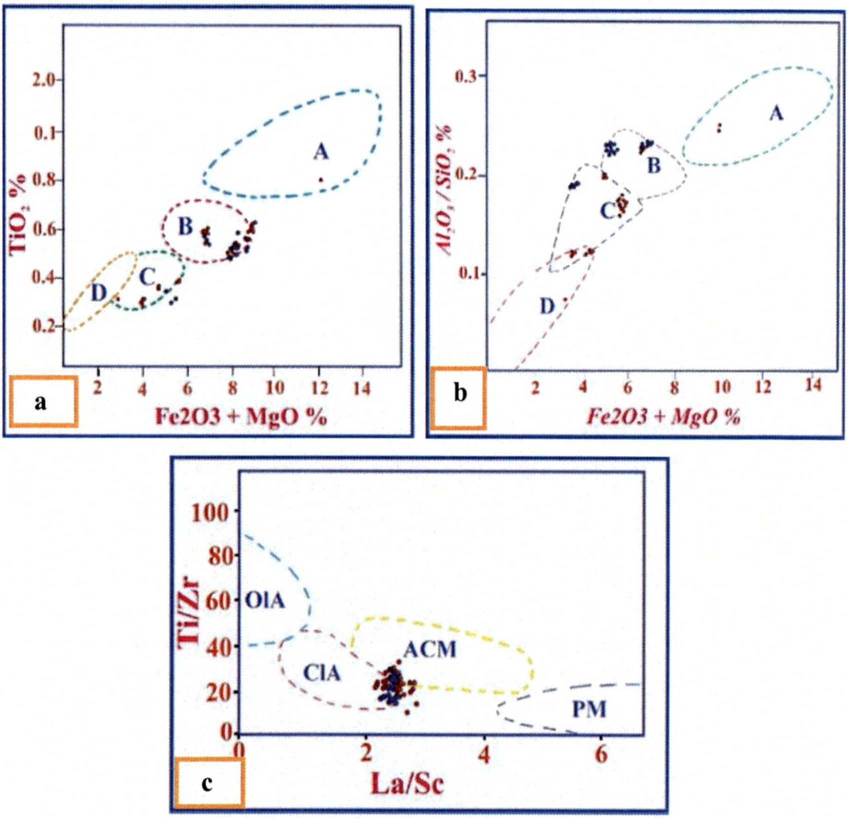

Fig. 4.15 Determining the tectonic position of sediments in Dezful (red dots) and Ramhormoz (blue dots). **a** Main element composition diagram on Bahatia diagram (Bhatia 1986) based on TiO_2 versus Fe_2O_3 + MgO. **b** Separator diagram based on AL_2O_3/SiO_2 ratio versus Fe_2O_3 + MgO, based on Bhatia (1986) diagram. **c** Draw the composition of sediment sub-elements of Dezful region on the tectonic positioning diagram based on diagram (Bhatia 1986). A = OIA: Oceanic island Arc, B = CIA: continental island Arc, C = ACM: active continental margin, D = PM: passive continental margin

4.2.4.6 Ratios of Main Elements in Relation to Source Stone

Detrital sediments of oceanic arch islands, derived from calc-alkaline andesites, are distinguished from other sedimentary samples by higher levels of Na_2O, AL_2O_3, TiO_2 and Fe_2O_3 and relatively low amounts of SiO_2 and K_2O (Gorokhov et al. 2019). Detrital sediments of continental arc islands, resulting from acidic and volcanic rocks, with higher SiO_2 and K_2O content and K_2O/Na_2O ratio (about 0.6) and lower Fe_2O_3 + MgO content than detached sediments of oceanic arch islands. They are separated (Heidari and Raheb 2020). Detrital sediments of active continental margins (anodic type and landslide basins) are mainly derived from the uplift. These sediments, which

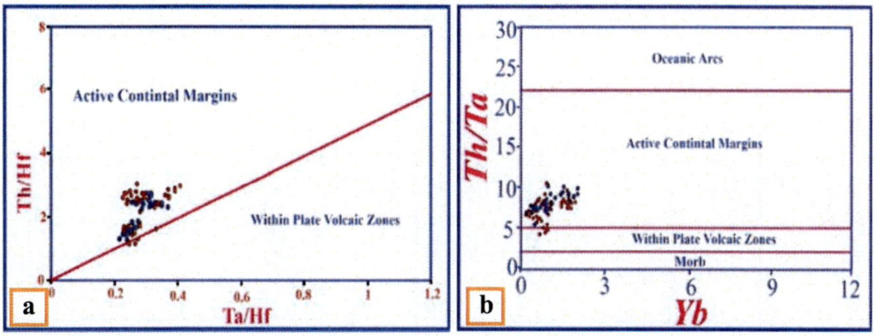

Fig. 4.16 Determining the tectonic position of sediments in Dezful (red dots) and Ramhormoz (blue dots) based on diagrams (Schandl and Gorton 2002). **a** Diagram of determining the tectonic environment based on Th/Hf to Ta/Hf in the study areas. **b** Diagram of determining the tectonic environment based on the ratio of Th/Ta to Yb in the study areas based on the diagram (Roser and Korsch 1986)

reflect the composition of the continental upper crust (UCC), have high amounts of SiO_2, K_2O with a ratio of K_2O/Na_2O (about 1) (Shahriar et al. 2020).

Inactive margin detrital sediments are generally rich in SiO_2 and free of Na_2O, AL_2O_3, TiO_2 and CaO (Heidari and Raheb 2020). The K_2O/Na_2O ratio in these sediments is more than one but may show many changes in composition, sometimes even similar in composition to the detrital sediments of active continental regions (Sharma et al. 2020). In both Dezful and Ramhormoz regions, due to the low amount of SiO_2 (an average of 24 for Dezful samples and an average of 20.1 for Ramhormoz samples) and low $Fe_2O_3 + MgO$ ratio, they are similar to the detrital sediments of the continental arch islands. However, considering the ratio of K_2O/Na_2O (about 0.9) and based on the total percentage of $Fe_2O_3 + MgO$, it makes them similar to the samples of the active continental margin and upper continental crust.

Determination of sediment source rock using main and secondary elements of felsic source rock containing higher values of Sr and Ba and lower amounts of Ni, Co, Cr, V than mafic source rock (Shahriar et al. 2020). If the abundance of Ni, Co, Cr, V along with MgO increases in the studied samples, it indicates a mafic origin for these sediments (Sahraeyan and Bahrami 2016). As can be seen in Fig. 4.17a, d, most of the samples are of felsic origin and some of the samples are of mafic origin, which indicates the transport of sediments from both sources to the mentioned areas (Fig. 4.17).

In this diagram, the amount of element V along with MgO shows a consistently increasing trend (Fig. 4.17a) that the increase in the percentage of V with MgO indicates the mafic origin of these sediments. In addition, the abundance of Ni along with MgO indicates the mafic origin for these sediments in the study areas (Fig. 4.17b). Also, the reciprocal increase of Co with MgO indicates the mafic origin for these sediments (Fig. 4.17c). Finally, as shown in this figure, element V along with the amount of MgO has an increasing trend that the frequency of V along with the

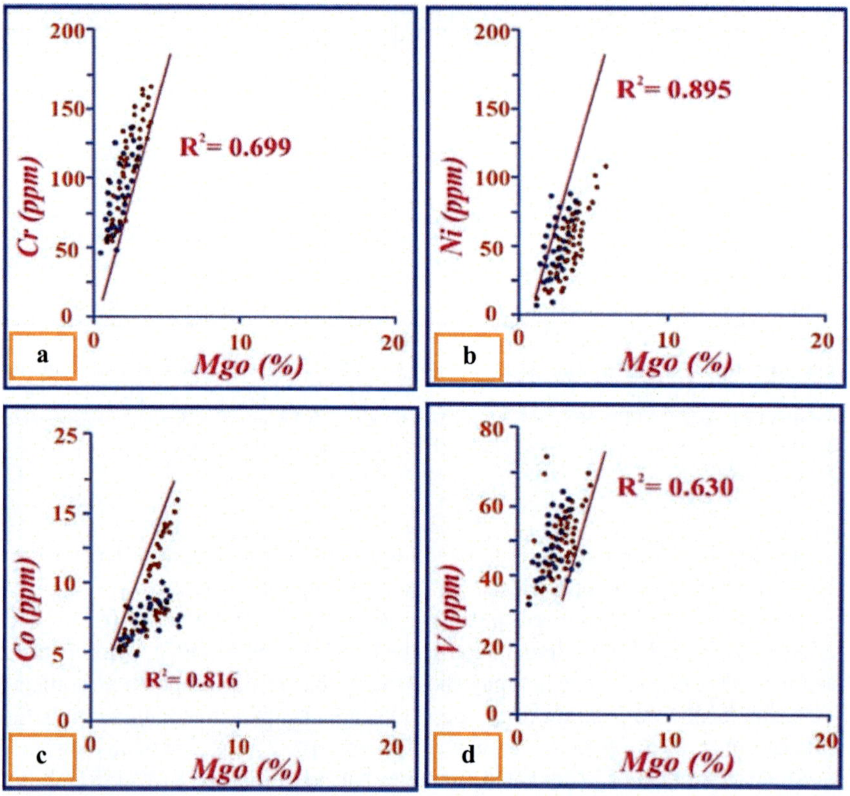

Fig. 4.17 Ratios of the main elements in relation to the tectonic position in Dezful (red points) and Ramhormoz (blue points) according to the diagram (Osae 2006). **a** The relationship between the percentage of magnesium oxide with the Cr sub-element according to the diagram. **b** The relationship between the percentage of magnesium oxide with the Ni element based on the diagram. **c** The relationship between the percentage of magnesium oxide with the CO element according to the diagram. **d** The relationship between the percentage of magnesium oxide with the V element

increase in the percentage of MgO indicates the mafic origin for these sediments (Fig. 4.17d).

The diagrams of Hayashi et al. (1997), obtained by plotting TiO_2 versus Zr, are used to distinguish between intermediate and felsic source rocks (Fig. 4.18a). Sub-elements such as Nb, Zr, Hf, Th, La are more abundant in acidic rocks and elements such as Ni, Cr, Co and Sc are more abundant in mafic rocks (Heidari and Raheb 2020). Cullers and Podkovyrov (2002) used these elements to separate the source rock (Fig. 4.18b). Cr/V versus Y/Ni and L a/Th versus Hf (Floyd and Leveridge 1987) are also used to identify the source rock (Fig. 4.18c).

As shown in Fig. 4.18a, the amount of TiO_2 in the Ramhormoz region varies between 0.1 to 0.5% and in the Dezful region between 0.2 and 0.8%, while the amount of Zr in the Ramhormoz region generally varies between 80 and 110 pp. M

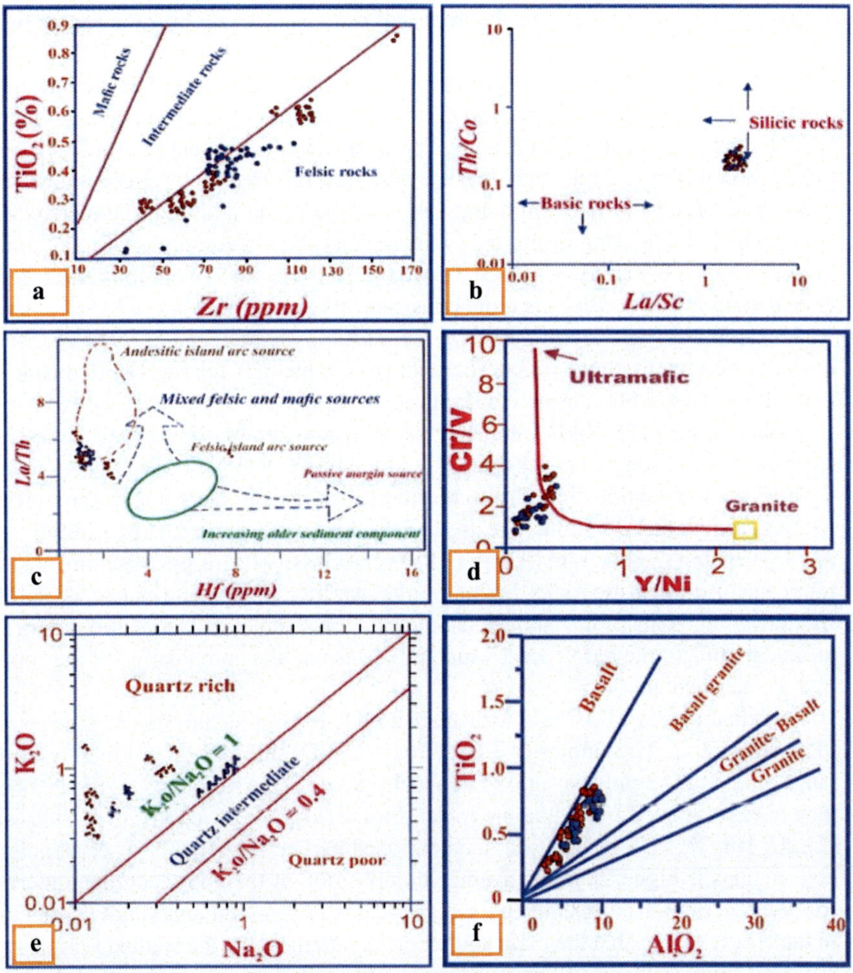

Fig. 4.18 Determining the source rock of studied sediments in Dezful study areas (red points) and Ramhormoz (blue points). **a** Mapping the values of the studied samples on the TiO_2 and Zr diagrams (Hayashi et al. 1997). **b** Combining the source rock of the studied samples on the Th/Co diagram versus La/Sc (Cullers and Podkovyrov 2002), **c** plot the values of the trace elements of the studied samples on the La/Th versus Hf diagram (Floyd and Leveridge 1987), **d** plot the values of the trace elements of the studied samples on the Cr/V diagram versus Y. **e** Determination of source composition based on Na_2O–K_2O diagram (Crook 1974) in studied sediments. **f** AL_2O_3 versus TiO_2 diagram to determine the origin of studied sediments based on the diagram (Amajor 1987)

is changing but this range is wider in the Dezful region and is from 40 to 130 ppm (Fig. 4.18a). According to this diagram, the composition of sediments in the Dezful region is generally of felsic origin, but in the Ramhormoz region, these sediments are probably of intermediate felsic origin. These samples are of siliceous origin in both Th/Co versus La/Sc diagrams (Fig. 4.18b) for both study areas.

In addition, La/Th versus Hf diagrams has been used to combine the source rocks of sediments in the study areas (Floyd and Leveridge 1987) (Fig. 4.18c). Based on this diagram, it can be concluded that the source rocks of the Felsic and Mafic interfaces for the sediments of Dezful and Ramhormoz can be suggested. The Hf value in both regions varies between 1 and 3 ppm. The tendency of sediments toward the felsic origin in both regions indicates that Hf has been released from zircon by erosion of the bow or spear (Al-Hashim and Corcoran 2020). In addition, the tendency of sediments to be of mafic origin may indicate that there were inputs of mafic rocks in both areas. This analysis is based on data from the Cr/V diagram versus Y/Ni. (Floyd and Leveridge 1987) is quite consistent (Fig. 4.18d).

According to some researchers, the main source of sediments is three groups of felsic, mafic and ultramafic rocks. The origin of sedimentary quartz is also considered in the felsic range (Sahraeyan and Bahrami 2016).

Based on the K_2O–Na_2O frequency percentage diagram, sediments can be classified into three categories: rich quartz, medium quartz and poor quartz (Fig. 4.18e). By drawing the data obtained from the studied samples in Dezful and Ramhormoz regions, sedimentary samples are located in the quartz-rich area. In addition, the AL_2O_3 versus TiO_2 diagram (Amajor 1987) can classify the origin of sediments into four categories: basaltic rocks, basalt granite, basalt granite and granite (Fig. 4.18f). By drawing the data of the studied samples, all the samples of the Dezful region are located in the granite basalt area, which indicates that the composition of the source rock is intermediate.

Examination of Fig. 4.18a–f, which are used to determine the nature of the source rock, shows that the samples of Ramhormoz and Dezful regions are mainly in the range of felsic to slightly basaltic source rocks. Some of the major and minor elements may have originated from alkaline rocks (Fe_2O_3 (MgO, Cr, Co, Ni)) or felsic rocks (Th, U, Nb, Ta, Hf). In addition, as mentioned earlier, the amount of carbonate in both regions is higher than the average composition of the upper continental crust, so the effect of high carbonate on the percentage of other elements is not unaffected. In fact, the reason is that the main and secondary elements in the studied samples are provided from rocks of different origins and the overlap of the amount of elements relative to each other on different diagrams has caused the samples to be placed in different ranges.

Validation of geochemical data of the studied samples by petrological studies conducted in different regions shows that the results obtained by petrographic results in the Sanandaj area of Sirjan and Urmia Dokhtar in terms of felsicity of most samples and mafic nature of some samples matches. In fact, extensive studies on the petrology of Sanandaj, Sirjan and Urmia Dokhtar zones indicate felsic and sometimes mafic origin in these areas, which can indicate the origin of the studied sediments in the mentioned areas. Among the studies that have been done on these areas, we can refer to petrological and geochemical studies in the Sanandaj area of Sirjan by Ranin et al. (2010) or studies conducted by Mohajel and Sahandi (1999) on the tectonic evolution of Sanandaj-Sirjan area in the northern half of the northwest and also the petrological studies carried out in the middle part of the Urmia Dokhtar magmatic strip by Goli et al. (2019) pointed out that in all the mentioned sources, the composition of rocks in these areas has been mainly felsic to mafic. In fact, studies show that the two identified sources are consistent with mafic masses with a combination of gabbro and diorite in the Sanandaj zone of Sirjan, and also the felicitous origin of these sediments with the felsic masses of SanandajSirjan and Urmia Dokhtar zone due to the mass. The granite sites of Boroujerd and Shahrekord are consistent and these results confirm the geochemical data obtained.

4.3 Mineralogy Studies

Mineralogy studies by X-ray diffraction (XRD) along with petrographic studies show very good information about the mineralogical properties of the studied sediments (Barnes et al. 2020).

In these studies, 12 samples from the Dezful region and 11 samples from the Ramhormoz region were analyzed by XRD machine in Sects. 4.1 of Ramhormoz and Dezful (Tables 4.10 and 4.11).

The data obtained from these studies show that chlorite and illite are the most abundant clay minerals in the studied deposits, especially in the Dezful region. In the Dezful region, calcite, quartz, feldspar and dolomite are among the non-clay minerals in these deposits, and in the Ramhormoz region, in addition to these minerals, gypsum is seen as another non-clay mineral. The following section examines the two main clay minerals observed in the laboratory results.

Table 4.10 Mineralogical results obtained from the study of samples from the Dezful region

No	D1	D2	D3	D4	D5	D6	D7	D8	D9	D10	D11	D12
XRD RESULTS	CALCITE+QUARTZ+FELDSPAR(minor)	CALCITE+QUARTZ+FELDSPAR(minor)	CALCITE+QUARTZ+DOLOMITE(minor)+FELDSPAR(minor)	CALCITE+QUARTZ+FELDSPAR+DOLOMITE	CALCITE+QUARTZ+DOLOMITE+CHLORITE+ILLITE+FELDSPAR(minor)	CALCITE+QUARTZ+FELDSPAR+CLAUMINERAL(minor)+DOLOMITE(minor)	CALCITE+QUARTZ+DOLOMITE++FELDSPAR(minor)+CLAUMINERAL(minor)	CALCITE+QUARTZ+DOLOMITE+FELDSPAR+CHLORITE+ILLITE	CALCITE+QUARTZ+CHLORITE+ILLITE+FELDSPAR(minor)+DOLOMITE(minor)	CALCITE+QUARTZ+FELDSPAR+DOLOMITE+CLAUMINERAL(minor)	CALCITE+QUARTZ+DOLOMITE	CALCITE+QUARTZ+FELDSPAR(minor)+DOLOMITE(minor)

Table 4.11 Mineralogical results obtained from the study of samples from the Ramhormoz region

No	R1	R2	R3	R4	R5	R6	R7	R8	R9	R10	R11
XRD RESULTS	CALCITE+QUARTZ+DOLOMETE(minor)+FELDSPAR(minor)+CLAYMINERAL(minor)	CALCITE+QUARTZ+FELDSPAR(minor)+DOLOMETE(minor)+CLAYMINERAL(minor)	CALCITE+QUARTZ+FELDSPAR+ILLITE DOLOMETE+CLAYMINERAL(minor)	CALCITE+QUARTZ+DOLOMETE+FELDSPAR + ILLITE (minor)	CALCITE+QUARTZ+DOLOMITE+FELDSPAR+CLAYMINERAL+GYPSUM	CALCITE+QUARTZ+DOLOMITE+FELDSPAR+CLAYMINERAL+GYPSUM	CALCITE+QUARTZ+DOLOMETE+GYPSUM+FELDSPAR+CLAYMINERAL(minor)	CALCITE+QUARTZ+FELDSPAR+DOLOMITE(minor)	CALCITE+QUARTZ+DOLOMETE+GYPSUM+FELDSPAR+CLAYMINERAL(minor)	CALCITE+QUARTZ+DOLOMETE+FELDSPAR+GYPSUM+CLAYMINERAL(minor)	CALCITE+QUARTZ+DOLOMETE+FELDSPAR+CLAYMINERAL+ GYPSUM (minor)

4.3.1 Illite

Ilite is formed in the source rocks of potassium ion suppliers. These ions may be obtained from the decomposition of minerals such as potassium feldspar and mica in the study areas. These ions can also be obtained from the alteration of volcanic rocks, pyroclastic and metamorphic rocks (Chamley 1989). The existence of such rocks in the areas of origin of the mentioned sediments (Sanandaj zone of Sirjan or Urmia Dokhtar) has been confirmed by petrographic studies, including the studies of Ranin et al. (2010), Mohajel and Sahandi (1999) and Goli et al. (2019) pointed out.

In addition, the presence of illite ores usually reflects the temperate climate conditions associated with chemical weathering in the area of origin (Oliveira et al. 2002). But chemical weathering during re-deposition cycles can also lead to illite formation (Mitra et al. 2018). Due to the hot-humid climatic conditions of the studied areas, the conditions of illite formation can be considered more in connection with the re-deposition cycles and leaching of sediments rich in feldspar and muscovite.

4.3.2 Chlorite

Chlorites are usually composed of alterations of classical volcanic particles such as biotite and amphibole and are a major component of lichens (Mitra et al. 2018). Chlorite can be obtained by alteration of feldspars. Feldspars are very unstable under normodynamic conditions similar to the surface of the earth and decompose rapidly under the influence of hot and humid climates and are converted to chlorite and other clay minerals by losing Ca, Ni, Si in many's. Are (Raghmi 2007).

Examination of XRD laboratory data in the Ramhormoz area indicates the absence of this mineral in this area (Table 4.11). Regarding the absence of this mineral, several reasons can be stated, but it seems that the absence of this mineral in this area is due to the fact that the source rock of these sediments, which is mainly in the Sanandaj area of Sirjan, is of the rock type. It is felsic that these types of rocks can not be a suitable source for the formation of minerals such as chlorite or smectite (Ranin et al. 2010). Therefore, it can be concluded that due to the low amount of this mineral in the studied samples, it was not possible to identify them by the XRD device used.

It should be noted that more accurate identification of clay minerals in the study areas requires the treatment of these samples with two methods of heat and saturation of ethylene glycol, while these results have been studied in general (Bulk) in the laboratory.

References

Abu M, Sunkari ED, Gurel G (2020) A paleocurrent analysis, petrographic, geochemical and statistical appraisal of Neoproterozoic siliciclastic sediments, NE Voltaian Basin, Ghana: a multidisciplinary approach to paleogeographic reconstruction. J Sedimentary Environ
Afarin M, Mahboubi A, Gargij M (2010) Geochemistry of mudstones and sandstones in the eastern region of Chabahar. In: Proceedings of the 15th conference of the geological society of Iran, 23 and 24 Azar 1390. Tarbiat Moallem University of Tehran, p 11
Al-Hashim MH, Corcoran PL (2020) Geochemistry study of Espanola formation, Bruce Mines—Elliot Lake area, Ontario, Canada: implications for provenance, paleo-weathering, and tectonic setting. Geosci J
Amajor LC (1987) Major and trace element geochemistry of Albian and Turronian shales from the Southern Benue trough, Nigeria. J Afr Earth Sci 6:633–641
Arvin A, Hajipour D, Heydari M (2016) A study of the trend of changes in some climatic elements in Chaharmahal and Bakhtiari province. Appl Res Geogr Sci 16(41):153–176
Barnes M, McLeod CL, Chappell C (2020) Characterizing the geogenic background of the Midwest: a detailed mineralogical and geochemical investigation of a glacial till in southwestern Ohio. Environ Earth Sci 79:159–167
Bhatia MR (1986) Trace element characteristics of greywakes and tectonic setting discrimination of sedimentary basins. Contrib Mineral Petrol 92:181–193
Blair TC, McPherson JG (1999) Grain size and textural classification of coarse sedimentary particles. J Sediment Res 69:6–19
Boggs SJ (2018) Principles of sedimentology and stratigraphy, 8th edn. Pearson Education, Upper Saddle River, New Jersey, pp 662–697
Chamley H (1989) Clay sedimentology. Springer-Verlage-Berlin, p 623
Cullers RL, Podkovyrov VN (2002) The source and origin of terrigenous sedimentary rocks in the Mesoproterozoic Ui group, southeastern Russia. Precambr Res 117:157–183
Dickinson WR (1985) Interpreting provenance relation from detrital modes of sandstones. In: Zuffa GG (ed) Provenance of arenites. Reidel, Dordrecht, pp 333–363
Floyd PA, Leveridge BE (1987) Tectonic environment of the Devonian Gramscatho basin, south Cornwall: framework mode and geochemical evidence from turbiditic sandstones. J Geol Soc 144:531–542
Folk E (1980) Petrography of sedimentary rocks. Hemphill Publishing Company, p 182
Ghosal S, Agrahari S, Banerjee S (2020) Geochemistry of the heavy mineral sands from the Garampeta to the Markandi beach, southern coast of Odisha, India: implications of high contents of REE and Radioelements attributed to placer monazite. J Earth Sci 129:152–167
Goli Z, Torabi Q, Maleki H (2019) Petrology of volcanic rocks in Dastgerdoo-Mullah Ahmad region (east of Isfahan), the middle part of the Urmia-Dokhtar magmatic strip. Kharazmi J Earth Sci 5(2):275–292
Gorokhov I, Zaitseva T, Kuznetsov AB, Ovchinnikova GV (2019) Isotope systematics and age of authigenic minerals in shales of the Upper Riphean Inzer Formation, South Urals. Strat Geol Correl 27(2):133–158
Grew E (2009) Geochemistry of sedimentary rocks. Mineral Soc Am 596–612
Hayashi KI, Fujisawa H, Holland HD, Ohmoto H (1997) Geochemistry of approximately 1.9 sedimentary rocks from northeastern Labrador, Canada. Geochim Cosmochim Acta 61:415–437
Heidari A, Raheb A (2020) Geochemical indices of soil development in arid to sub-humid climosequence of Central Iran. J Sci 17:165–179
Herron MM (1988) Geochemical classification of terrigenous sands and shales from core or log data, J Sediment Res 58:820–829
Hessler AM, Lowe DM (2006) Weathering and sediment generation in the Archean: an integrated study of the evolution of siliciclastic sedimentary rocks of the 3.2 Ga Moodies Group, Barbeton Greenstone Belt, South Africa. Percambrian Res 151:185–210
Jalilian A (1396) Advanced sedimentary environments. Payame Noor University Press, p 347

Jin ZF, Li J, Cao S, Wang JY (2006) Geochemistry of Daihai Lake sediments, Inner Mongolia, north China: implications for provenance, sedimentary sorting and catchment weathering. Geomorphology 80:147–163

Krishnamurthy P (2020) Rare metal (RM) and Rare Earth Element (REE) resources: world scenario with special reference to India. J Geol Soc India 95:465–474

Mitra R, Chakrabarti G, Shome D (2018) Geochemistry of the Palaeo-Mesoproterozoic Tadpatri shales, Cuddapah Basin, India: implications on provenance, paleoweathering and paleoredox conditions. Acta Geochimica 37(5):715–733

Mohajel M, Sahandi M (1999) Tectonic evolution of Sanandaj Sirjan zone in the northwestern half and introduction of new sub-zones in it. Geol Surv Miner Explor, Q J Earth Sci 31–32

Musawi Harami R (2004) Sedimentology, 9th edn. Astan Quds Razavi Printing and Publishing Institute, p 474

Nesbitt HW, Young GM (1982) Early Proterozoic climates and plate motions inferred from major element chemistry of lutites. Nature 299:715–717

Oreshkina TV, Aleksandrova GN, Lyapunov SM (2020) Micropaleontological and lithogeochemical characteristic of the Turtas Formation, Western Siberia. Stratigr Geol Correl 28:311–329

Paikaray S, Banerjee S, Mukherji S (2008) Geochemistary of shales from the Paleoproterozoic to Neoproterozoic Vindhyan Supergroup: implications on provenance, tectonics and paleoweathering. J Asian Earth Sci 32:34–48

Pettijohn FJ, Potter PE, Siever R (1987) Sand and sandstone. Springer, Berlin, p 553

Raghmi M (2007) Introduction to clay minerals. Gorgan University of Agricultural Sciences, p 262

Ranin A, Sepahi AA, Moein-Vaziri H and Aliani F (2010) Petrology and geochemistry of the plutonic complexes of the Marivan area, Sanandaj-Sirjan zone (In Farsi with English abstract). J Petrol, Isfahan University, Iran 2:43–60

Roser BP, Korsch RJ (1986) Provenance signatures sandstone- mudstone suites determined using discriminate function analysis of major element data. Chem Geol 67:119–139

Sahraeyan M, Bahrami M (2016) Geochemistry of sandstones from the Aghajari Formation, Folded Zagros Zone, southwestern Iran: implication for paleoweathering condition, provenance, and rectonic setting. Int J Basic Appl Sci 4:390–407

Schandl ES, Gorton MP (2002) Application of high field strength elements to discriminate tectonic setting in VMS environment. Econ Geol 97(3):629–642

Shahriar MS, Ameen SMM, Hossain MS (2020) Revealing the basement in Barapukuria: a geochemical study of a Gondwana Coal Basin basement from Northwest Bangladesh. J Geol Soc India 95:571–586

Sharma RP, Raja P, Bhaskar BP (2020) Pedogenesis and mineralogy of alluvial soils from semi-arid southeastern Part of Rajasthan in Aravalli Range, India. J Geol Soc India 95:59–66

Smirnov PV, Konstantinov AO, Aleksandrova GN, Kuzmina OB, Shurygin BN (2019) New data on the lithology of coastal facies of the Turtas Formation (Upper Oligocene, Southwestern Siberia). Earth Sci 475(6):868–871

Sorokin AA, Smirnov YV, Smirnova YN (2017) Geochemical features and sources of clastic material in terrigenous deposits of the northeastern flank of the South Mongolia-Khingan orogenic belt. Stratigr Geol Correl 25:146–166

Suttner LJ, Dutta PJ (1986) Alluvial sandstone composition and palaeoclimate: framework mineralogy. J Sedimeantary Petrol 56:329–345

Chapter 5
Tectonic and Morphotectonic Studies of Alluvial Fans

Abstract Tectonics is one of the branches of geology that studies the deformation of the earth's crust due to stresses and strains during different geological periods (Pazzaglia 2017). The main purpose of this science is to explain the geometry and origin of geological structures and structures such as anticlines, faults, joints and synclines. On the other hand, tectonics studies the internal structure of the earth, and how mountain ranges, oceans, earthquakes and other surface events occur (David et al. 2005). Since familiarity with the tectonic and morphotectonic properties of the studied areas is the basic principle of any type of geochemical and sedimentological studies, in this chapter, we have tried to study these characteristics in the study area as a suitable model for this type of study in Introduced worldwide.

Keywords Structural geology · Faults · Foldings · Morphotectonic

5.1 The Concept and Scope of Structural and Morphotectonic Geology

Structural geologists use the term construction for a part of the earth that has been deformed or deformed by force acting on the surface or within the earth's crust (Mazur et al. 2020). The structure consists of a geometric arrangement of plates, lines, surfaces, rock masses, etc. (Punturo et al. 2014). The shape and orientation of this arrangement is a reflection of the performance between the deformation force and the primary rock mass (Ortolano et al. 2020).

Structural geology discusses the structure of rocks, the deformation and the geometric position of structures formed in rocks (Ma 2020). This branch of earth science, then, deliberately discusses the geometric shape and physical characteristics of various earth structures, such as layers, folds, faults, salt domes, lines, etc., and the mechanism by which they are formed (Patel et al. 2020).

It is very important to study the geometry of the structure (determining the relationship between the components of the rock or the structure before and after the

S. Pourmorad and A. Mohanty, *Alluvial Fans in Southern Iran*, Advances in Geographical and Environmental Sciences, https://doi.org/10.1007/978-981-19-2045-5_5

deformation), so to achieve this goal the process of deformation or change of appearance of the rock assembly (deformation) must be studied (Mukherjee 2013). The metamorphic process is responsible for the formation of geological structures (Patel et al. 2020).

5.2 Tectonic Analysis and How to Study Structures in Geology

For the analysis of tectonic structures, based on the accuracy and scale studied, geoscientists have proposed various classifications, which are referred to here by the method proposed by the German Academy of Sciences.

- Global tectonics: Tectonic analysis of geological structures on a global scale at the level of continental or oceanic blocks (Punturo et al. 2014).
- Macro-tectonics: Accurate collection and description of plate elements or linear elements in the body of a rock mass the size of internal rock masses or metamorphism (Passchier and Trouw 1996).
- Microtectonics: The study and determination of symmetric and geometric relationships as well as the relationship between recrystallization and deformation by studying thin sections of rocks (Passchier and Trouw 1996).

Since the knowledge of tectonic and morphotectonic properties is one of the most basic and necessary information in alluvial fan studies, in this chapter, the aim is to study some important tectonic features of the studied areas. These case studies can be used as a suitable model for studying alluvial fans around the world It is noteworthy that tectonic and morphotectonic studies have a very wide range of studies, but since the main purpose of this chapter is geochemical studies and sedimentology of alluvial fans, only some of the basic and necessary cases in this type of studies are mentioned. Relevant sources can be used for further studies.

In summary, in this chapter, the aim is to first provide complete information about the tectonics of the study areas of southwestern Iran (Dezful and Ramhormoz regions) separately and then to study the role of tectonics in the formation, displacement and sedimentation of particles by morphotectonic studies of these areas. In these cones, fans should be addressed.

5.2.1 Tectonic and Morphotectonic Properties of Ramhormoz Region

5.2.1.1 Faults

The study of faults plays a very important role in sedimentary studies, especially in alluvial fan sediments, due to their importance in conducting and infiltrating water, creating secondary porosity and forming springs and ponds (Beygi et al. 2016). The study of faults can provide comprehensive information about sediment transport and sedimentation in addition to the initial information of the sedimentary basin (Gawthorpe et al. 2018). Faults in the area include thrust and strike-slip faults. The main fault in the region is the Ramhormoz thrust fault, which has a northwest-southeast direction, and its slope is to the northeast. Right-slip and normal faults are sub-faults and have more frequency.

Ramhormoz Fault

Ramhormoz fault with a length of more than 100 km is located in southwestern Iran and adjacent to Ramhormoz city (Aghanbati 2016). This fault is located in the famous depression of southwestern Iran called the Dezful depression and is active from a tectonic point of view (Ghanbarian et al. 2021). The main extension of this fault is northwest-southeast and is located in the active folded orogenic belt of Zagros (Hatzfeld et al. 2011). According to Cooper (2007) the Zagros fold and thrust belt (Zagros FTB) are an approximately 1,800-km (1,100 mi) long zone of deformed crustal rocks, formed in the foreland of the collision between the Arabian Plate and the Eurasian Plate.

Ramhormoz fault in the southern part of the Persian anticline has caused the eviction of the Gachsaran Formation (Evaporitic sediments with Lower Miocene age) on Bakhtiari Formation (alluvial sediments with Pliocene age) (Alavi 2004). In fact according to Pourmorad et al. (2021), the Persian anticline is a semi-symmetrical anticline in the north of Ramhormoz city, which has a northwest-southeast direction and is located at the southwestern edge of the Zagros orogenic belt and in the northern part of the Dezful depression. The southwest edge of this anticline is inverted (Figs. 5.1 and 5.2).

The southwestern edge of this anticline from the east of Ramhormoz city to the northwest of this city has been completely removed or covered by Quaternary sediments (Aghanbati 2016). The rate of vertical movement and uplift on the fault from its southeastern part to the northwest gradually increases so that the southeastern part of the Gachsaran Formation is driven by the Mishan Formation (Carbonate sediments with Middle Miocene age) and to the northwest is located on younger units (Alavi 2004). Underground data from oil wells and seismic data show that the level of Ramhormorz drift separation is located in the studied sheet at the peak of the

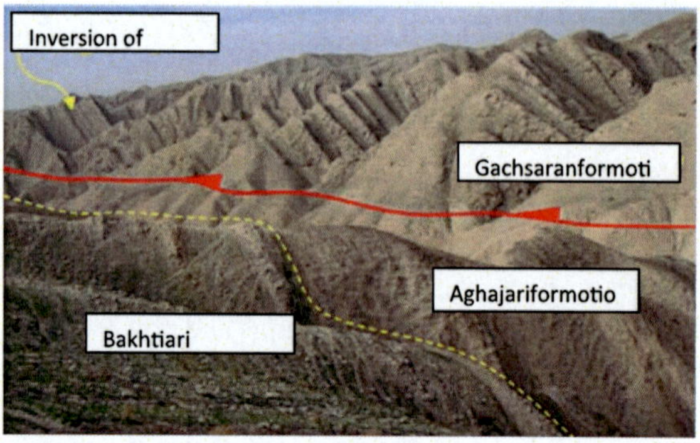

Fig. 5.1 Drift of Gachsaran Formation on Lahbari section and reversal of layers in the southwest edge of Parsis anticline

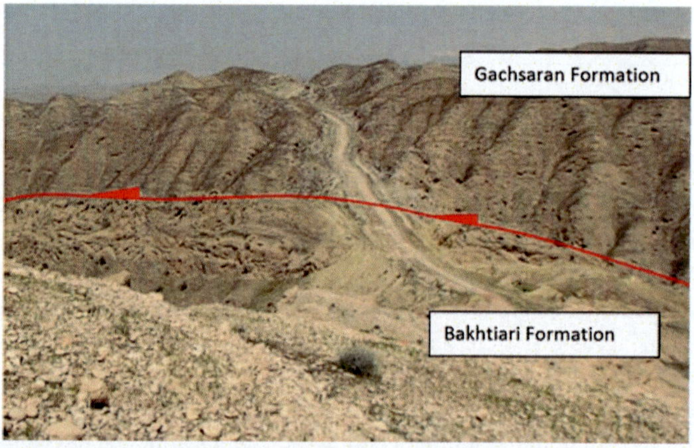

Fig. 5.2 Drift of Gachsaran Formation on Bakhtiari Conglomerate due to Ramhormoz Fault—Southwest Edge of Parsi Anticline, Northeast View, Southwest of Iran

Persian anticline and is the result of a cut between Asmari and Gachsaran Formations (Aghanbati 2016).

5.2.1.2 Foldings

Foldings are important in hydrogeological studies due to their role in water conduction and karst appearance (if the climate is humid and limestone) (Yassaghi and Marone 2019). According to Shamszadeh et al. (2016), the movement and cycle of water along the perpendicular to the fold layers is limited and increases in parallel.

Fig. 5.3 An example of a parasitic fold in the Gachsaran Formation as the source of some of the alluvial fans studied in southwestern Iran

In the study area, folded structures include anticlines and synclines that are consistent with the structural trend of the Zagros fold and thrust belt (Zagros FTB) (Aghanbati 2016).

In fact, Zagros FTB has a northwest-southeast structural direction that stretches from the Taurus Mountains in southeastern Turkey to the Minab Fault in the east of the Strait of Hormuz in southern Iran (Alavi 2004).

The study area consists of several Chinese, which is called the main China of the Persian China region. In addition, several sub-folds are observed in the Gachsaran Evaporation Formation to the Early Miocene (Fig. 5.3).

5.2.1.3 Geomorphology of the Studied Areas

Understanding the geomorphology of the study areas is one of the essential points in identifying and studying alluvial fan deposits. In fact, without knowing the geomorphology of the region, the results of many other studies will have low accuracy.

An example of geomorphological studies for a model in this type of study is the study of some geomorphological items related to this type of study in the Dezful and Ramhormoz regions in southwestern Iran In this type of study, it is necessary to study the shape of the area, altitudes, depressions, waterways, lithology and other geomorphological information about the area.

According to Aghanbati (2016), in terms of geomorphological studies, it can be said that the Ramhormoz region in the southwest of Ebran has different appearance variability due to the structure of the folded Zagros. Also according to Pourmorad et al. (2021), in this region, the north and northeast are the heights of the Gachsaran

Evaporation Formation (San Miocene) with an approximate height of 464 m above sea level. These heights are considered as the central part of the Persian anticline (as previously discussed this anticline) (Aghanbati 2016). Lithologically, these elevations mainly include soft white to light gray gypsum, light anhydrite and white to light gray halites with gray and red marls (Pourmorad et al. 2021).

In addition to the heights of the Gachsaran Formation in the north of this region, low-lying hills are observed in the northwestern parts of the region, which are mainly due to sedimentary streams and river flows that have been transferred from the heights of the Gachsaran Formation to this region. Sediments in this section mainly include fine-grained to medium-grained sandstone, silt, clay and evaporates (white gypsum, anhydrite and white to gray salt) (Alavi 2004).

The third zoomorphological complication of the region is the low and flat sediments of the region. These sediments, which are mainly composed of silt and clay, form the end parts of the large Ramhormoz alluvial fan, which have settled due to the reduction of transport energy intensity (mainly river flows). The area of this part gradually increases from the northeast of the region to the southwest of the region (Pourmorad et al. 2021).

Waterways in most areas are in the form of tree branches or dendrites. In rock outcrops, their bed is narrow, but in the plains, it is wide and covered with loose and isolated sediments (Pourmorad et al. 2021). The Madani, Kopal and Shifa rivers are the main rivers in the region. Alluvial barracks are formed around the Madani River, Shifa and other waterways and are seen at different altitudes (Fig. 5.4).

Fig. 5.4 Satellite image of the ester showing the study area, which shows the position of the Persian anticline in relation to the Ramhormoz plain

5.2.1.4 Morphotectonic Properties

In the morphotectonic analysis of the region, the use of the geomorphic index as a primary and rapid tool to estimate the relative changes of tectonic activities has been expanded (Karasozen et al. 2019). This information plays a very important role in the study of alluvial fans (Collinson 1996).

According Gao et al. (2016) geomorphic indices and morphometric parameters are calculated to investigate the role of neotectonic activities, as it acts as a controlling factor in the development of landforms in the tectonically active terrains.

To check the morphotectonic properties of the studied areas from various indicators such as Mountain Front Sinusity Index or Mountain Front Sinuosity (SMF), Drainage Basin Asymmetry (Af), Basin Shape Index (Basin) Shape: BS) and River Sinuosity Index (S) are used (Pourmord et al. 2021) 7-2, which are discussed in the following section.

Mountain Front Sinusitis Index or SMF Mountain Forehead Maze

The sinusoidal index expresses the conflict between the erosive forces that try to erode the mountain front on the one hand and the tectonic force that causes the smooth and direct force on the other hand, if the erosive forces prevail. The high rate of mountain front maze (SMF) is shown, and if the tectonic forces are active, the SMF number will be lower (Yassaghi and Marone 2019). This index is calculated from the following equation:

$$\mathrm{SMF} = \mathrm{LMF}/\mathrm{LS}$$

SMF = mountain front maze

Length Mountine front (LMF) = The length of the mountain front along Paykooh

Length straight (LS) = The length of a straight line on a mountain front.

The SMF is between 1 and 1.6 for highly tectonic active areas, between 1.4 and 3 for moderately active areas, and from about 1.8 to more than 5 for passively tectonically active mountain fronts (Collinson 1996).

In Ramhormoz study area in southwestern Iran, according to the existing structure, it was divided into two parts. In the first part, the length of the mountain front along Paykooh (LMF) was 39.4 km and the length of the tangent line on the mountain front (LS) was equal to 36.6 km that were calculated, therefore:

$$\mathrm{SMF} = 39.4/36.6$$

$$\mathrm{SMF} = 1.07$$

For the second part, the mountain front along Paykooh (LMF) was equal to 83 km and the length of the tangent line on the mountain front (LS) was equal to 65 km.

$$\text{SMF} = 83/65$$
$$\text{SMF} = 1.27$$

Watershed Asymmetry Index (Af)

This index is defined as follows:

$$\text{AF} = 100(\text{Ar}/\text{At})$$

In this equation AF, the index of waterway asymmetry, Ar, the area of the basin containing the sub-drainage on the right bank of the main waterway (in kilometers per square meter), the area of the basins containing the sub-drainage on the left and right bank of the main waterway in km per square meter (Shirahama et al. 2019).

In areas with active uplift due to the uplifted topography on one side of the area and subsidence on the other, the length and area of these waterways on both sides of the uplift of the area will be more than the same length and area on the opposite side. If the numerical values of this index are around 50, it indicates the symmetry of the sub-drainages to the main waterway and as a result there is no skew in the area due to uplift. Numerical values greater than 50 or less than 50 will indicate the uplift performance to the right and left of the main canal (Collinson 1996).

The asymmetry index is a method for detecting skewness due to tectonic activity in drainage basins (Shirahama et al. 2019). This index allows proving the lateral tilt of a basin relative to the main course of the river (Shayan 2003). The Af index in the Ramhormoz study area was calculated as follows (Fig. 5.5):

$$\text{Right Ar} = 8.17\ \text{km}^2$$
$$\text{Left At} = 14.98\ \text{km}^2$$
$$\text{Af} = (8.17/14.98)100 = 54.53$$

Basin Shape Index (BS)

The basin shape index is an ideal criterion in the study of tectonically active basins from inactive basins so that technically active basins are elongated (Shirahama et al. 2019). The shape of the basin over time with the amount of stopping the ridge gradually forms a circle, in other words, it can be understood that the shape is one of the characteristics of basins that are located in mountainous areas with high tectonic activities. Over time, the shape of the basin rises in the region and approaches the circle (Karasozen et al. 2019). In active tectonic areas, the width of the catchment is narrower because in such areas the river force is used to dig and erode its bed, while tectonic stability causes the river to have the opportunity to widen its bed (Shirahama et al. 2019). The shape ratio of the basin is obtained from the following equation:

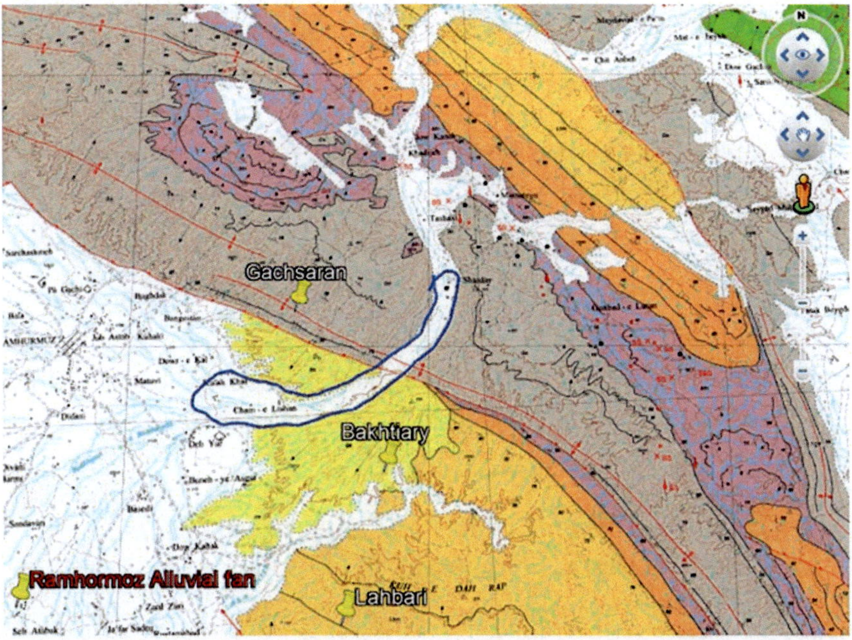

Fig. 5.5 Location of Ala River (in blue) to study the Af index in the 1:10,000 Haftgol map in southwestern Iran (Setudehnia and Perry 1966)

$$BS = BI/BW$$

BS: Basin shape ratio

BI: The size of the length of the basin from the end of the water divider to the outlet of the basin

BW: The width of the basin in the widest part.

Basins characterized by large amounts of BS, i.e. more than 4, are tectonically dormant. When BS is between 3 and 4, the basin is in Class 2 in terms of tectonic activity, and values less than 3 indicate circular basins that are inactive (Shirahama et al. 2019).

The results of studying this index in the Ramhormoz region are as follows:

$$Bi = 114\ \text{Km}$$

$$BW = 26\ \text{Km}$$

$$BS = 4.38$$

As can be seen, the area in question has a value of more than 4 and represents the active tectonic basin. Therefore, tectonic activities are among the main factors in the movement of sediments in the alluvial fan.

River Maze Index (S)

This index is obtained as a practical measure of the ratio of river length (C) to the ratio of valley length to a straight line (V). In this index, the higher the value of this index, the amount of S (river maze index) indicates the active tectonics in the region (Dorn 2009).

$$C = 114 \text{ KM}$$
$$V = 1.51 \text{ KM}$$
$$S = 75.49$$

As can be seen, the high amount of river maze indicates the tectonic activity of the area.

5.2.2 *Another Example of Tectonic and Morphotectonic Studies (Dezful Study Area)*

The study area is a part of the simple-folded Zagros in terms of rock facies and structural appearance (Alavi 2004). The outcrops in the area include Neogene stratigraphic units belonging to the Fars group. The stratigraphic units of the Fars Group include Miocene deposits of the Bakhtiari Formation and finally Quaternary deposits in river channels. Young Quaternary deposits are widespread in the area, often at the foot of heights and at the core of synchronization, and eventually as regoliths on structures.

5.2.2.1 Folding

The folds in the study area are located in a relatively high plain, but in the north, east and southeast of the Zagros Mountains. In the northeast of the study area, there is the Chenareh anticline, which has many folds, faults and structural complications (Fig. 5.6).

5.2.2.2 Dezful Study Area Faults

There are several faults and fractures in the northern and southern ridges of Chenareh anticline at different lengths along the axis of China, which are discussed in the following section.

Fig. 5.6 The folds of the Bidroubeh anticline in southwestern Iran

5.2.2.3 Dezful Fault

This quaternary fault with a length of more than 200 km and a northwest-southeast direction with a slope to the northeast is of the listric thrust type, which gradually decreases in-depth at a depth. In the study area, this fault is mostly underground and no evidence of a specific surface level is observed, and the existence of this fault has been confirmed based on the underground data of the National Company for Southern Oilfields.

5.2.2.4 Mountain Front Fault (MFF)

This fault actually limits the simple folded Zagros belt to the south and southwest and is structurally very important seismic topography and can be thought of as a major thrust fault, which is buried.

Geological evidence, such as the condition of the upper parts of the Asmari Formation, stratigraphic characteristics, seismic and drilled wells, indicates that there is a vertical displacement along the mountain front fault that is more than 6 km. As a result of vertical movements along this fault, the southwestern edge of the creased belt rises along this fault and thus asymmetric surface folds that are also located on the front have been obtained (Aghanbati 2016).

Specifically, the mountain ridge fault consists of several complex and separate drift sections that have a length between 15 and 115 km and their total length is more than 1350 km in Iran (Aghanbati 2016). These fault fragments are separated at depth, along with the corresponding asymmetric folds (which are on the surface) and are separated from each other by specific manifestations and steps that can be

Fig. 5.7 Mountain front fault in the study area of Dezful, southwestern Iran

seen in morphotectonic topographic features. These parts form two wide arcs in the Fars region (south-center plate) and Lorestan (northwest of Zagros) so that they have right and left ladder arrangements and are located on the east and west sides of these two arc patterns (Aghanbati 2016).

The mountain front fault has been displaced by the Kazerun-Borazjan fault for at least 140 km in the right direction (Alavi 2004). Studies on seismic activity along various parts of the mountain edge drift fault show that the studied earthquakes are located in the cut of folded areas. In this case, there were several earthquakes from 1052 to 1991, the magnitude of which varies between 5.5 and 6.8 (Aghanbati 2016). These earthquakes appear to form in areas where there is no Pisangi fault and then propagate around this area. Investigation of the mechanism of earthquakes along the mountain front fault indicates that the fault is almost a pure thrust type and the obtained surfaces are parallel to the trend of geological buildings in the area and the mountain front fault (Fig. 5.7).

5.2.2.5 Joints and Gaps

Joints are one of the most common fractures observed in karst environments (Karasozen et al. 2019). Because the presence of gravity water is also possible in cracks with a diameter of 1–2 microns, the joints in the ground can be water passageways from the surface to the depths (Sarkarinejad and Azizi 2008). Karst shapes such as Karen and Shaft are formed along the joints (Tatar et al. 2004). Tensile joints are of special importance in karst formation and development (Alavi 2004).

These types of joints have more openings than shear types (Karasozen et al. 2019). Fractures, called regular joints, are plates in which stress along them causes localized adhesion of the stone. The plane surface of fractures is a level of weakness (discontinuity) in the rock (Tatar et al. 2004). Although the definition of a fault makes sense despite the displacement at the fracture level, in practice in some cases it is very difficult to correctly distinguish joints and fractures from faults. Although fractures are divided into different types such as joints, shear zones, cracks and veins, but in this section, only the joints in the Chenareh anticline are examined. From a hydrogeological point of view, it is important to identify tensile and shear joints because tensile joints have more openings and create more hydraulically guided zones than shear joints (Teixell et al. 2018).

In general, the shear joints on the limestone surface of the Asmari Formation in southwestern Iran show a parallel displacement parallel to the joint surface and usually appear in conjugate clusters at an angle of about 60 degrees to each other. The shear joints in the anticline, unlike the tensile joints, have a small opening and are closed. Genetically, the joints in the anticline are formed by tectonic factors resulting from the forces forming the fold. The joints in the edges of the anticline that are perpendicular to the axis of the fold are tensile joints because when the limestone layers are pressed along the perpendicular to the axis of the anticline, they form the anticline that are stretched along the axis of the anticline. In addition to tensile joints perpendicular to the Chinese axis, there are joints parallel to the axial surface of the anticline that can be considered as release joints (Lashgari and Derikvand 2020).

These joints are mainly around the ridge of the anticline. Shear joints that form a 60-degree angle with each other can be seen in the edges around the anticline. In fact, when the layers are pressed along a perpendicular to the axis of the anticline, in addition to producing the anticline, two sets of shear joints are formed. In the vicinity of faults, several joints are often observed that have a shear origin. These joints are mainly filled with feathers on both sides of the fault line and at an acute angle to it (Fig. 5.8).

5.2.2.6 Morphotectonic Properties of Dezful Region

The most important morphotectonic properties calculated in the Andimeshk region are as follows:

Mountain Front Sinusitis Index or SMF Mountain Forehead Maze

As mentioned, this index is calculated from the following equation:

$$\mathrm{SMF} = \mathrm{LMF}/\mathrm{LS}$$

SMF = mountain front maze
LMF = length of mountain front along Paykooh
LS = straight line length on the mountain front.

Fig. 5.8 Close view of the joint harvesting station on the southern edge of the Chenareh anticline in southwestern Iran

SMF is between 1 and 1.6 for highly tectonic active areas, between 1.4 and 3 for moderately active areas, and from about 1.8 to more than 5 for passively tectonically active mountain fronts (Sujit and Krishnendu 2019).

In this region, according to the existing structure, it was divided into two parts. In the first part, the length of the mountain front along Paykooh (LMF) was equal to 65.61 km and the length of the tangent line on the mountain front (LS) was equal to 34.3. Kilometers were calculated accordingly:

$$\text{SMF} = 65.61/34.3$$
$$\text{SMF} = 1.91$$

For the second part, the length of the mountain front along Paykooh (LMF) was equal to 40.38 km and the length of the tangent line on the mountain front (LS) was equal to 19.8 km:

$$\text{SMF} = 40.38/19.18$$
$$\text{SMF} = 2.1$$

As can be seen, the value of the mountain front sinusoidal index in the region in question is between 1.4 and 3, which indicates a region with moderate and weak tectonic activity and confirms the lower tectonic role of sediment yield in the region. In the discussion section, the reasons are discussed.

Watershed Asymmetry Index (Af)

As mentioned, this index is defined as follows:

$$\mathrm{Af} = (\mathrm{Ar}/\mathrm{At}) * 100$$
$$\mathrm{Ar} = 0.20\ \mathrm{km}^2$$
$$\mathrm{At} = 0.53\ \mathrm{km}^2$$
$$\mathrm{Af} = (0.11/0.32)100 = 37.73$$

As can be seen in the area in question, the Af index is less than 50, which indicates the uplift on the west side of the area and thus more sediment in this area.

Basin Shape Index (BS)

The results of the basin shape (BS) survey in the Dezful study area include the following results:

$$\mathrm{BS} = \mathrm{BI}/\mathrm{BW}$$
$$\mathrm{Bi} = 123\ \mathrm{Km}$$
$$\mathrm{BW} = 37\ \mathrm{Km}$$
$$\mathrm{BS} = 3.32$$

As can be seen, the area in question has a value between 3 and 4 and represents a basin with moderate to low tectonic activity.

River Maze Index

As mentioned, this index is calculated from the following equation:

$$\mathrm{S} = \mathrm{C}/\mathrm{V}$$

C: The length of the river
V: The length of the valley in a straight line.
The results in the study area are as follows:

$$\mathrm{C} = 3.43\ \mathrm{KM}$$
$$\mathrm{V} = 2.97\ \mathrm{KM}$$
$$\mathrm{S} = 3.43/2.97$$
$$\mathrm{S} = 1.15$$

This index indicates the weak and moderate tectonic activity of the region and confirms the low tectonic role of sediment yield.

5.3 Discussion

As can be seen in the table below, the results of morphotectonic studies show that the Ramhormoz region has an active to semi-active tectonic position. Undoubtedly, this active tectonic position is one of the main factors in sediment transport in the alluvial fans (Table 5.1).

Also, as can be seen from the table below, the results of morphotectonic studies show that the Dezful region has a moderately weak tectonic position (Table 5.2).

In the following section, an attempt is made to investigate the technical conditions of the study area to measure the morphotectonic results obtained.

Examination of stratigraphic, sedimentological and tectonic evidence shows that metamorphism in the Zagros is not a continuous process but has occurred in different stages. The first compressive movements in the China Belt and the Zagros drift were in the Late Cretaceous, following the subduction of the young Tethys oceanic crust and the initial collision of the Iranian and Arabian plates (Alavi 2004). As the forehead converges, the deformity migrates to the southwest. Deformation in the simple folded belt begins at the end of the Eocene, and the fold belt expands episodically with the addition of a small number of active folds at a time (Hessami et al. 2001).

Wrinkling has started in the Dezful river in Miocene following the continent–continent collision between the Iranian and Arabian plates. The first stage of folding coincides with the deposition of the upper parts of the Gachsaran Formation, which is accompanied by the formation of growth layers in these deposits. The second stage of folding is characterized by the presence of growth layers in the Upper Aghajari that were deposited during the Upper Miocene. This stage has ended with the extensive deposition of the Bakhtiari Formation. The third stage of deformation that forms the current appearance of the Zagros is characterized by folding and faulting in the Bakhtiari Formation. The conglomerate was formed at the same time as the Pliocene orogeny with the facies of pig back basins and rubble from the Persian group erosion at this stage (Sherkati and Letouzey 2004).

Morphotectonics such as the establishment of alluvial barracks and folding of Quaternary deposits and tectonic seismic evidence such as earthquakes indicate the continuation of tectonic efforts in the region (Abiat et al. 2017). These efforts are the result of the continued convergence of the Arabian and Eurasian plates, which occurs in an approximately north–south direction at a rate of about 25–35 mm per year

Table 5.1 Results of morphotectonic properties of the Ramhormoz region

SMF	Af	BS	S
1.27	54.53	4.38	75.49

Table 5.2 Results of morphotectonic properties of the Dezful region, southwest of Iran

SMF	Af	BS	S
2.1	37.73	3.32	1.15

(Talebian and Jackson 2004). 10–12 mm of shortening resulting from this convergence occurs in the China Belt and the Zagros Drift and the rest in the active belts of KopehDagh—Alborz—Talesh and Absheron—Balkans (Talebian and Jackson 2004). A small part of the shortening also occurs on right-slip faults and important drifts in central Iran. In the China belt and the Zagros drift, shortening and deformation are not evenly distributed and occur mostly around active faults such as the mountain frontal fault and the main young fault of the Zagros (Sherkati and Letouzey 2004).

In the study area in the Ramhormoz area, the presence of the Ramhormoz drift fault has an important role in intensifying tectonic activity and the resulting shortening and uplift. This uplift, by intensifying erosion in the supra-wall areas of drift, has increased the sediment yield of rivers and large waterways and the formation of alluvial fans in the front of the fault and its sub-wall. However, in the Dezful region, due to the fact that the active fault structure is not visible on the surface, sedimentation of rivers is mostly the result of regional uplift and erosion processes, and tectonic activities play a lesser role in their formation.

References

Abiat A, Ajdari A, Judaki M, DarvishiKhatunabadi J (2017) Study and separation of quaternary sedimentary environments in Khuzestan plain. J Adv Appl Geol 7(3):49–64

Aghanbati A (2016) Geology of Iran, 1st edn. Geological Survey and Mineral Exploration Organization, Tehran, 708 p

Alavi M (2004) Regional stratigraphy of the Zagros fold-thrust belt of Iran and its proforeland evolution. Am J Sci 304:73–97

Beygi S, Nadimi A, Safaei H (2016) Tectonic history of seismogenic fault structures in Central Iran. J Geosci 61:127–144

Collinson JD (1996) Alluvial sediments. In: Reading HG (ed) Sedimentary environments: processes, facies and stratigraphy. Blackwell, Oxford, pp 37–82

Cooper M (2007) Structural style and hydrocarbon prospectivity in fold and thrust belts. Deformation of the continental crust: the legacy of Mike Coward. Special Publications, vol 272. Geological Society, London, pp 447–472

David D, Pollard R, Fletcher (2005) Fundamentals of structural geology. Cambridge University Press, New York, 500 p

Dorn RI (2009) The role of climatic change in alluvial fan development. In: Parsons AJ, Abrahams AD (eds) Geomorphology of desert environments. Springer, Dordrecht, pp 723–742

Gao M, Zeilinger G, Xu X, Tan X, Wang Q, Hao M (2016) Active tectonics evaluation from geomorphic indices for the central and the southern Longmenshan range on the Eastern Tibetan Plateau, China. Tectonics 35(Issue 8)

Gawthorpe RL, Leeder MR, Kranis H, Skourtsos E, Andrews JE, Henstra GA, Stamatakis M (2018) Tectono-sedimentary evolution of the Plio-Pleistocene Corinth rift Greece. Basin Res 30(3):448–479

Ghanbarian MA, Yassaghi A, Derakhshani R (2021) Detecting a sinistral transpressional deformation belt in the Zagros. Geosciences 11(6):226. https://doi.org/10.3390/geosciences11060226

Hatzfeld D, Authemayou C, Van der Beek P, Bellier O, Lavé J, Oveisi B, Tatar M, Tavakoli F, Walpersdorf A, Yamini-Fard F (2011) The kinematics of the Zagros Mountains (Iran). In: Leturmy

P, Robin C (eds) Tectonic and stratigraphic evolution of Zagros and Makran during the Mesozoic-Cenozoic. Special Publications, vol 330. Geological Society, London, pp 19–42

Hessami K, Koyi HA, Talbot CJ, Tabasi H, Shabanian E (2001) Progressive unconformities within an evolving foreland fold–thrust belt, Zagros Mountains. J Geol Soc Lond 158:969–981

Karasozen E, Nissen E, Bergman EA, Ghods A (2019) Seismotectonics of the Zagros (Iran) from orogeny wide, calibrated earthquake relocations. J Geophys Res Solid Earth 124:919–929

Lashgari A, Derikvand B (2020) Folding style of the Jarik Anticline NE Dezful embayment of Zagros Belt, the role of thickness changing of the folded layers and syntectonic deposition. J Pet Sci Eng 189:136–152

Ma Y (2020) Regional tectonic setting and prototype basin evolution. In: Marine oil and gas exploration in China. Environmental Earth Science, online

Mazur S, Aleksandrowski P, Gągała L (2020) Late Palaeozoic strike-slip tectonics versus oroclinal bending at the SW outskirts of Baltica: case of the Variscan belt's eastern end in Poland. Int J Earth Sci (Geol Rundsch) 109:1133–1160

Mukherjee S (2013) Deformation microstructures in rocks. Springer Geochemistry/Mineralogy, Berlin, p 111

Ortolano G, Visalli R, Fazio E (2020) Tectono-metamorphic evolution of the Calabria continental lower crust: the case of the SilaPiccola Massif. Int J Earth Sci (Geol Rundsch) 109:1295–1319

Passchier CW, Trouw RAJ (1996) Microtectonics. Springer, Berlin, pp 196–212

Patel NK, Pati P, Verma AK (2020) Seismicity around the Mahendragarh-Dehradun basement fault in the western Ganga plain, India: a neotectonic perspective. Int J Earth Sci (Geol Rundsch) 109:689–706

Pazzaglia J (2017) Tectonic geomorphology of the Red Rock Fault, insights into segmentation and landscape evolution of a developing range front normal fault. J Struct Geol 27:1925–1939

Pourmorad S, Harami R, Solgi A, Ale Ali M (2021) Sedimentological, geochemical and hydrogeochemichal studies of alluvial fans for mineral environment purposes. Lithol Miner Resour 56:89–112

Punturo R, Cirrincione R, Fazio E, Fiannacca P, Kern H, Mengel K, Ortolano G, Pezzino A (2014) Microstructural, compositional and petrophysical properties of mylonitic granodiorites from an extensional shear zone (Rhodope Core complex, Greece). Geol Mag 151:1051–1071

Sarkarinejad K, Azizi A (2008) Slip partitioning and inclined dextral trenspression along the Zagros thrust system, Iran. J Struct Geol 30:116–136

Setudehnia A, Perry JT (1966) Haftkel geological compilation map 1:100000. Iranian Oil Company

Shamszadeh A, Alawi A, Wali Nejad M, Tavakoli Yerki M (2016) Wrinkle style and structural complexities in the Central Lorestan Sub-Zone. J Oil Explor Prod 144:48–53

Shayan S (2003) Geomorphological characteristics of alluvial fans of Gamasiab basin. Res, Geogr 46:99–113

Sherkati S, Letouzey J (2004) Variation of structural style and basin evolution in the central Zagros (Izeh zone and Dezful embayment), Iran. Mar Pet Geol 21(5):535–555

Shirahama Y, Miyairi Y, He H, Fu B, Echigo T, Yokoyama Y, Ikeda Y (2019) Late quaternary evolution of the Kumkol Basin at the northeastern margin of the Tibetan Plateau revealed by tectonic geomorphology and the analysis of in situ cosmogenic nuclides. Geomorphology 329:224–247

Sujit D, Krishnendu G (2019) Morphotectonic analysis of the Sali River basin, Bankura district, West Bengal. Arab J Geosci 12(7):244–258

Talebian M, Jackson JA (2004) A reappraisal of earthquake focal mechanisms and active shortening in the Zagros mountains of Iran. Geophys J Int 156:506–526

Tatar M, Hatzfeld D, Ghafori-Ashtiany M (2004) Tectonics of the Central Zagros (Iran) deduced from microearthquake seismicity. Geophys J Int 156:255–266

Teixell A, Labaume P, Ayarza P, Espurt N, de Saint Blanquat M, Lagabrielle Y (2018) Crustal structure and evolution of the Pyrenean-Cantabrian belt: a review and new interpretations from recent concepts and data. Tectonophysics 724–725:146–170

Yassaghi A, Marone C (2019) The relationship between fault zone structure and frictional heterogeneity, insight from faults in the High Zagros. Tectonophysics 762:109–120

Chapter 6
Environmental Geological Studies

Abstract Careful environmental studies, including pollution, environmental hazards, flooding and other factors, are an integral part of any careful study of alluvial fan sediments (Pourmorad et al. 2021). In this chapter, the aim is to study the various factors affecting the environment of these areas in terms of various environmental pollutants (soil pollution and groundwater quality assessment) and various hazards (including seismicity and flooding). What makes these studies important along with the studies of alluvial fan sediments in these areas is the very close and direct relationship between these sediments and the increase and intensification of hazards and pollution in these areas. In fact, the type of sediment, the slope of alluvial fans and the intensity of debris currents can cause human catastrophes in these areas along with high-risk earth activities such as earthquakes, surface runoff and landslides. On the other hand, the type of lithology of the constituent formations can also have a direct impact on groundwater quality. In fact, the main purpose of writing this chapter was to go one step beyond the science of pure sedimentology and by applying it, to provide this information to the safety of human society in the study areas, to achieve this goal, the missing link It will be between pure studies and applied studies and the strength of these studies. In this chapter, first, environmental pollution in terms of the presence of heavy elements and rare earth elements and groundwater quality assessment of these areas are studied, and in the next stage, geological hazards in terms of seismicity and flooding in the studied areas are investigated.

Keywords Environmental pollution · Heavy metalsm · Igeo · CF · Groundwater chemistry · Environmental hazards

6.1 Environmental Pollution

In this section, first, the soil contamination of the target areas in terms of the presence of heavy metals is investigated, and in the next stage, the quality assessment of groundwater resources in the area of the studied alluvial fans is investigated.

S. Pourmorad and A. Mohanty, *Alluvial Fans in Southern Iran*, Advances in Geographical and Environmental Sciences, https://doi.org/10.1007/978-981-19-2045-5_6

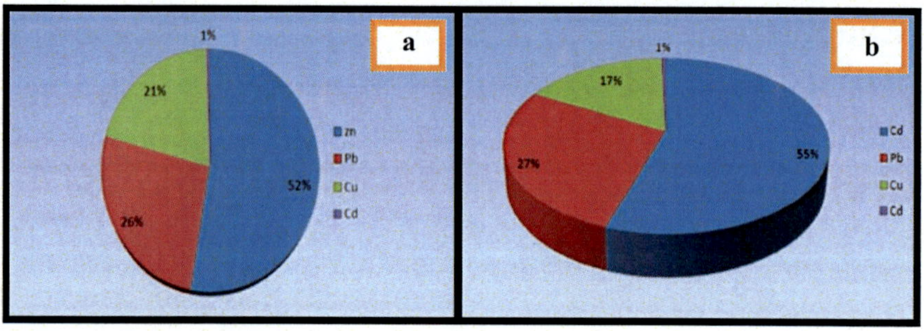

Fig. 6.1 Frequency of heavy metals in the studied areas in terms of percentage. **a** Dezful study area, **b** Ramhormoz study area in southwestern Iran

6.1.1 Study of Heavy Metals

Today, soil pollution with heavy metals is considered as one of the most important environmental pollutants (Lin et al. 2020). These metals are classified as the most dangerous group of pollutants in terms of toxicity and stability, and because of their soil pollution properties, toxicity, long shelf life and their accumulation in living tissues, it is important. There are many ecological and biological aspects (Park et al. 2020). In fact, heavy metals refer to a group of metals that have a specific gravity of more than 6 g per cubic meter or an atomic mass of more than 50 (Lin et al. 2020).

Studies in southwestern Iran have been conducted to investigate the distribution of heavy metals in the sediments of this region and have been used as a model for studies around the world. According to these studies, in Dezful region, the average frequencies of heavy metals include Zn with a frequency of 39 ppm, Pb with a frequency of 19.5 ppm, Cu with a frequency of 15.2 ppm and Cd with a frequency of 0.34 ppm (Fig. 6.1a). In Ramhormoz region, the average frequencies of these metals are Zn with a frequency of 37.8 ppm, Pb with a frequency of 19 ppm, Cu with a frequency of 11.5 ppm and Cd with a frequency of 0.33 ppm, respectively (Fig. 6.1b).

6.1.2 Study of Environmental Indicators

There are different methods for calculating the degree of enrichment of sediments by heavy metals. Indicators discussed in this study include the following.

6.1.2.1 Igeo Geoaccumulation Index

This index is a common method for estimating sediment contamination with heavy metals, in which the concentration of heavy metals in sediments is obtained relative to the concentration of the metal background.

This index, presented by Muller (1969), is based on the following relationship:

In this relation, Cn is the measured concentration of the element in the sample and Bn is the concentration of the same element in the background sample, which is obtained from the relation opposite to this concentration:

$$\text{x} + \text{s}\,(\text{middle} + \text{standard deviation}).$$

In Muller (1969) equation, a coefficient of 1.5 is used to correct the effects of soil parent materials, natural fluctuations and very few changes caused by human activities. Based on this index, soils are classified into seven different groups as follows (Table 6.1).

The results of the study of this index for Dezful and Ramhormoz regions are presented in Tables 6.2 and 6.3.

Table 6.1 Grading of sediment contamination level according to Müller cumulative geochemical index (Muller 1969)

Number obtained for Igeo	Category Igeo	Sediment contamination status
0	0	Completely uninfected
0–1	1	Non-contaminated to moderate contamination
1–2	2	Moderate pollution
2–3	3	Moderate to severe pollution
3–4	4	Severe pollution
4–5	5	Severe to very severe pollution
5–6	6	Very severe pollution

Table 6.2 Results of Igeo index study in Dezful study area in southwestern Iran

	Cn	Bn	log2
Cu	15.2	18.7	−0.8839
Zn	39.1	52.7	−1.0156
pb	15	30	−1.585
Ni	48	62.4	−0.9635
V	48.8	62.5	−0.9419

Table 6.3 Results of Igeo index study in Ramhormoz study area in southwestern Iran

	Cn	Bn	Cn/1.5bn	log2
Cu	11.5	13.3	0.576441	−0.79475
Zn	37.8	50.2	1.894737	0.921997
pb	15	30	0.75188	−0.41143
Ni	48.7	66.3	2.441103	1.287533
V	44.8	56.8	2.245614	1.16711

6.1.2.2 Pollution Factor (CF)

Contamination factor is one of the common indicators to determine the level of soil contamination, based on which the amount of elements can be measured relative to its normal value and the amount of soil contamination can be determined. This factor is calculated according to the following relation for the studied elements (Liu et al. 2016).

$$CF = (C\,\text{metal})/(C\,\text{background}).$$

In this equation, the pollution index is obtained from the ratio of the concentration of each metal to the amount of natural ground concentration of that metal. The Hakanson classification (Hakanson 1980) is also used for the pollution factor to assess heavy metal pollution (Table 6.4).

The results of the pollution factor in the pollution factor (CF) method are presented for some of the studied samples in Dezful (Table 6.5) and Ramhormoz (Table 6.6).

6.1.3 *Discussion*

Interpretation of data from environmental studies is a major step in this type of study. In fact, by collecting and analyzing all the data, many important expectations can be achieved in this type of study. It is recommended that all analyzes be obtained with great care and from the overall conclusion of all data because one or two results may not reflect all the environmental properties of the area.

Table 6.4 Evaluation of heavy metal pollution level based on pollution factor (Hakanson 1980)

Severity of pollution	Amplitude of CF changes
Low pollution	1
Moderate pollution	1–3
High pollution	3–6
Severely infected	6≤

Table 6.5 Results obtained in Dezful region regarding the amount of CF in this region

NO	Cu	Zn	Pb	Ni	V
D1	0.78	0.48	0.50	0.45	0.62
D2	1.08	0.41	0.50	0.34	0.50
D3	0.67	0.58	0.50	0.57	0.64
D4	0.48	0.37	0.50	0.43	0.45
D5	0.66	0.54	0.50	0.45	0.52
D6	0.98	0.98	0.50	1.31	1.02
D7	0.67	1.32	0.50	0.53	0.68
D8	0.80	0.81	0.50	1.07	0.95
D9	1.29	1.17	0.50	1.69	1.57
10	0.73	0.77	0.50	0.85	0.86

Table 6.6 Results obtained in Ramhormoz region regarding the amount of CF in this region

NO	Cu	Zn	Pb	Ni	V
R1	0.811	0.472	0.5	0.423	0.631
R2	0.541	0.375	0.5	0.339	0.650
R3	0.801	1.090	0.5	0.671	0.960
R4	1.234	1.155	0.5	1.175	1.486
R5	0.773	0.631	0.5	0.754	1.021
R6	0.961	0.782	0.5	0.794	1.052
R7	0.872	0.706	0.5	0.843	1.098
R9	0.992	0.813	0.5	0.974	1.234
R10	0.785	0.752	0.5	0.642	0.871

As can be seen from the data obtained from the average pollution caused by Cu, Zn, Pb, Ni, V elements in both Ramhormoz and Dezful regions, in Dezful region the pollution caused by all the mentioned heavy metals is less than zero and in other words has Pollution is low but in Ramhormoz region in terms of Zn, Ni and v elements have moderate pollution.

The results of the contamination factor in the pollution factor (CF) method for the samples studied in Dezful and Ramhormoz regions are somewhat similar to the results of the calculation of the contamination method by the Müller method (Muller 1969) and indicate that in Dezful region relatively low pollution, but in Ramhormoz region, pollution caused by vanadium (v) and to some extent Ni and Zn can be observed. According to studies, the mentioned pollutants in Ramhormoz area are probably due to the presence of drilling fluids or oil leakage through joints and cracks in the rock cover to the upper parts of the soil in Mamatin oil field near Ramhormoz city. The mentioned environment in the environments around the rigs being drilled and penetrating into the soil causes pollution of natural resources and

Fig. 6.2 Existence of oil slicks in the water of Rudzard river due to Mamatin oil field (one of the important oil fields in southwestern Iran)

especially groundwater in these areas, while in Dezful region due to the absence of these conditions we see much less pollution than these elements (Fig. 6.2).

6.1.4 Groundwater Chemistry

Groundwater is one of the most important sources of human drinking water. The water side is salty. Therefore, by examining the quality of these resources and its continuous monitoring, the necessary knowledge can be applied in order to perform optimal management of water resources protection and prevention of future damages. Therefore, research on the chemical quality of groundwater is very important in determining environmental pollutants that are related to the improper disposal of various wastewater.

6.1.4.1 PH

pH represents the concentration of H + ions, ie hydrogen ions in water (Asare-Donkor and Adimado 2020). There is a numerical range from 0 to 14 for pH, where pH = 7 indicates neutrality, and less than this value, towards 0, indicates an acidic solution, and above 7, towards 14, indicates a solution (Wang et al. 2020). The lower the pH of the water, the more acidic the water tends to be and therefore the more it tends to corrode. The higher the pH, the more the water tends to be alkaline and precipitate (Asare-Donkor and Adimado 2020). According to the studies, the average

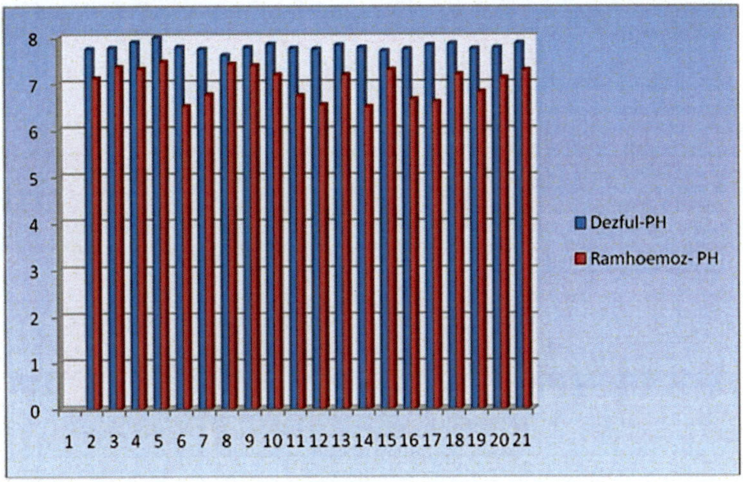

Fig. 6.3 Comparison of percentage of pH changes in blue samples of Dezful (blue) and Ramhormoz (red)

pH in Dezful region is 7.78 and in Ramhormoz region is 7, which is in the range of American Environmental Standard (EPA) and Iran (6.5–8.5) (Fig. 6.3).

6.1.4.2 Water Hardness (TH)

Hard water is water that contains mineral salts such as bicarbonate compounds, calcium ions and magnesium (Ding et al. 2020). The degree of hardness of water is determined by the amount of calcium and magnesium in it. The severity rating of water hardness is not a standard and varies in different parts of the world (Wang et al. 2020).

The severity of water hardness depends on the surface of the water flow in the surface and underground. Waters in calcareous areas are harder than waters in granite or sandy areas (Guevara et al. 2020). The hardness of water also changes over time. Water hardness usually decreases in the rainy season and increases in the dry season, and sometimes in rainy and wet seasons such as caves (in carbonate areas) (Esmeray and Gokcekli 2020).

Changes in water hardness in Dezful region are between 1.5 and 3.4 mg/l in change and the average is 2.5 mg/l. In Ramhormoz region, these changes are between 2.7 and 29.5 mg/l with an average of 12.6 mg/l. The above changes indicate that the water hardness of Dezful is lower than the world standard (4.5 mg/l) and the water hardness in Ramhormoz region is higher than this standard.

As can be seen, water hardness in Ramhormoz region is much higher than Dezful region, which is due to the fact that in Ramhormoz region Gachsaran Formation is an important and effective formation in the region and causes much higher hardness than regional samples. Dezful has become the source of Bakhtiari and Aghajari

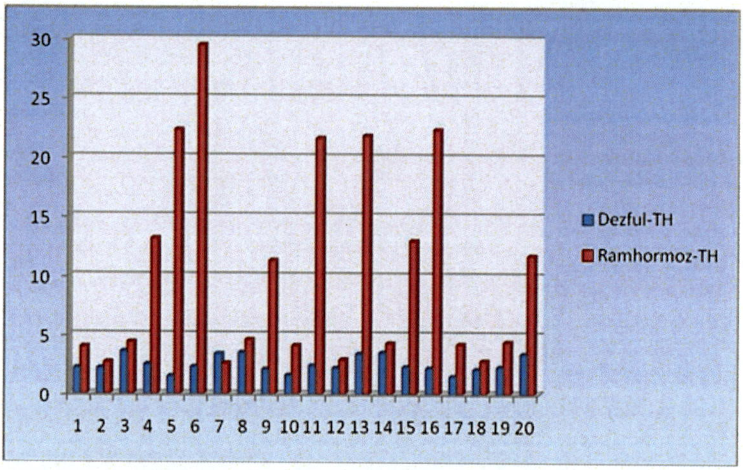

Fig. 6.4 Comparison of percentage changes of TH in blue samples of Dezful (red) and Ramhormoz (blue) in southwestern Iran

formations. This difficulty is more justifiable (Fig. 6.4). It should be noted that the samples were harvested in autumn when the intensity of rainfall was not heavy and undoubtedly with increasing rainfall intensity in both areas, the amount of TH in them decreases.

6.1.4.3 Electrical Conductivity (EC)

One simple way to determine the concentration of water-soluble solutes is to measure electrical conductivity (Jung 2020). Distilled water or pure water is almost not a conductor of electricity, but if it is present in the water of soluble salts, it conducts electricity. The more solutes dissolved in the water, the higher the electrical conductivity; in other words, its electrical resistance decreases (Gharahi and Zamani 2020).

The amount of electrical conductivity in Dezful region varies between 0.28 (mho/cm) and 0.65 (mho/cm) and with an average (0.40 mho/cm), which indicates the low electrical conductivity of water samples in this region, in other words, high electrical resistance of water samples in this area (Fig. 6.5). Also, samples of Ramhormoz region have electrical conductivity between 0.4 (mho/cm) and 6.7 (mho/cm) and with an average of 2.4 (mho/cm). According to the obtained data, it can be concluded that the electrical conductivity in aqueous samples of Ramhormoz region is high due to the presence of Gachsaran Formation and consequently the presence of salt, which is the main factor in increasing electrical conductivity (Fig. 6.5).

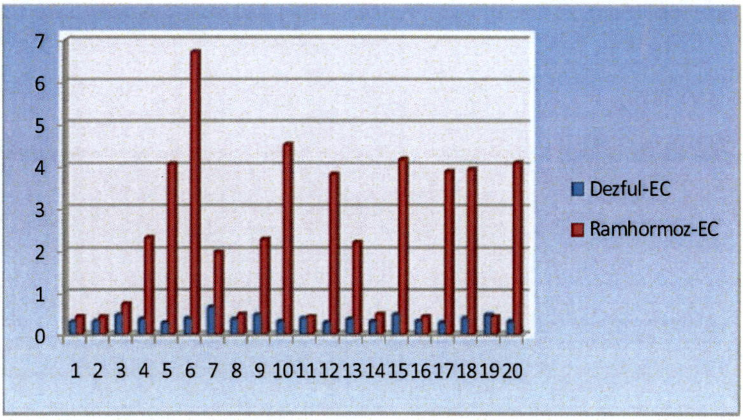

Fig. 6.5 Comparison of EC change percentage in blue samples of Dezful (blue) and Ramhormoz (red) regions in southwestern Iran

6.1.4.4 Sulfate (SO4)

The presence of sulfate in drinking and agricultural waters can cause problems such as undesirable taste in water, diarrhea, decreased detergent cleaning power, etc. (Gharahi and Zamani 2020). Therefore, the lower the percentage of sulfate in the aqueous sample, the higher the quality.

The amount of sulfate in the samples of Dezful region, except for one station, is very small and immeasurable, which indicates its desirable quality. But in Ramhormoz region, the amount of sulfate changes varies between 562 and 1377 mg/l. Comparison of sulfate ion concentrations in water samples with world water standards shows that the amount of sulfate in the waters of the region is generally higher than the global standard value of 960 mg/l. Higher concentrations of sulfate cause nucleic acid and as a result lead to poisoning of plants in the region (Jung 2020).

6.1.4.5 Potassium (K)

Potassium is found in a variety of minerals, which may be dissolved in water through weathering processes (Gharahi and Zamani 2020). Rivers generally contain about 2.3 PPM of potassium (Cellone et al. 2020). This difference is mainly caused by the high concentration of potassium in oceanic basalts (Kumar et al. 2020). Granites are rich in calcium up to 2.5% potassium (Liu et al. 2020).

According to studies, the changes in potassium in aqueous samples of Dezful region are between 0.1 and 1.8 and have an average of 0.6 ppm. In Ramhormoz region, these changes are between 0.9 and 4.8 and with an average of 3 ppm in change (Fig. 6.6).

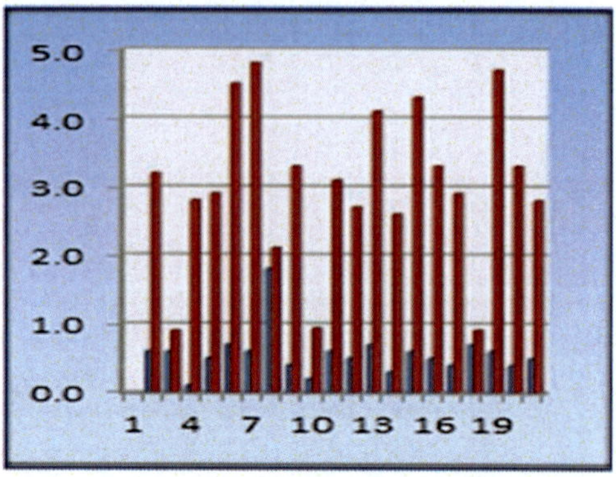

Fig. 6.6 Comparison of potassium percentage in two regions of Dezful (blue) and Ramhormoz (red)

6.1.4.6 Sodium (Na) and Calcium (Ca)

The amount of sodium in the water samples of Dezful region varies between 4.8 and 23.4 (PPM) with an average of 7.6PPM. These changes in Ramhormoz region change between 5.3 and 520 ppm and with an average of 190 ppm. Also, the amount of calcium in Dezful region is between 0.08 and 0.8 ppm with an average of 0.5, which changes in Ramhormoz region between 0.5 and 8.6 and with an average of 3.1.

6.1.4.7 Discussion of Hydrological Results

According to studies, the pH of the water of the studied wells in Ramhormoz and Dezful regions is normal and is in the range of 6.5 to 7.9, which is the standard for drinking water and agriculture of the US Environmental Protection Agency (EPA) and Iran (6.5–8.5) is not higher. Therefore, it does not impose restrictions on various uses such as drinking, irrigation and agriculture. It is noteworthy that the concentration of hydrogen ions in pure water sources is very important for various uses because it affects the taste, corrosion, efficiency of refining processes and industrial use (Esmeray and Gokcekli 2020).

Comparison of the amount of different elements in aqueous samples of Dezful and Ramhormoz areas shows that in Ramhormoz area the amount of potassium, sodium, calcium and sulfate elements is significantly higher than Dezful area. This significant difference in the two mentioned areas can be attributed to the existence of Gachsaran Formation in Ramhormoz region and as a result of high water hardness in this formation. In fact, the solubility of salt and gypsum as well as the erodibility of sediments in this formation is a limiting factor of passing water resources and causes undesirable and salinity of water and increases the amount of hardness and other elements in this region while in Dezful region The following conditions do not rule out the absence of the above formation.

The results show a very high salinity of water samples in Ramhormoz large alluvial fan compared to Dezful region. These results are consistent with the available evidence, including abundant salt deposits in the Ramhormoz region and vice versa, especially in the distal and central parts of the alluvial fans of Andimeshk and Dezful regions.

6.2 Environmental Hazards

Considering the importance and role of alluvial fans in the intensification of environmental hazards in this section, first seismic hazards and other related natural hazardous factors in Dezful and Ramhormoz regions are studied as a model and in the next section The study of flooding will be considered due to its importance and high role in environmental hazards in these areas.

6.2.1 Seismicity Studies and Seismic Related Factors in the Studied Areas

One of the important and related cases for identifying the transport and displacement of sediments in alluvial fans is recognizing the seismicity of the area. In areas with high seismicity, this seismicity can cause the sediments to move more than normal in the upstream areas and create a lot of natural hazards for the residents of these areas.

For this type of study, there are three types of analytical, probabilistic and statistical studies. In the analytical method, by studying the tectonics of the region and identifying different seismic sources (especially faults), the parameters of the earth's motion and its response rate according to the damping relationships that correlate with the desired parameter in terms of magnitude and distance. Is evaluated (Mehdizadeh and Kinematic 2019).

In the probabilistic method, seismic hazard analysis by determining the probabilistic models of seismic sources (point, linear or regional) and by considering the seismic parameters and using appropriate damping relations, the most motion parameters of Earth and its response rate are determined (Abd El-Aal et al. 2020).

The statistical method uses the statistical methods provided by the US Nuclear Regulatory Commission for Nuclear Power Plants to calculate the maximum acceleration and response rate of ground motion.

Gutenberg–Richter Introductory Method (G-R)

The Gutenberg–Richter basic distribution function was introduced in 1956. One of the best ways to study the seismicity of a region is to use the information of earthquakes that have occurred in that region over the past years and centuries (Mehdizadeh and Kinematic 2019). A study by Gutenberg–Richter found that there is a linear relationship between the frequency and intensity of earthquakes that occur in each region, which can be plotted based on these two variables. In this relation, the cumulative frequency of earthquakes N (M) is linearly attributed to magnitude (M) considering the simple relation below.

$$\log \mathrm{N(M)} = \mathrm{a} - \mathrm{bM}.$$

In the above relation N is the number of earthquakes. The above equation generally includes moderate earthquakes.

Even by using this simple relation, it is possible to obtain a seismic relationship and its related coefficients (coefficients a, b) by performing a suitable classification for the magnitude of earthquakes. However, due to the limitations of this relationship, on the one hand, it is not possible to use historical earthquakes and on the other hand, due to the lack of upper and lower limits for magnitude, the results are only a function of the mathematical nature of the relationship and are not compatible with seismic realities. Therefore, the results are relatively acceptable only for intervals when the seismic information is relatively homogeneous and continuous (instrumental earthquakes recorded in the twentieth century) and in a limited range of magnitude (7.5–4 M).

In addition to the linear relationships that exist between the frequency and intensity of earthquakes in each region, a non-linear relationship can also be considered for that region, which generally includes low to high intensity earthquakes. Some scientists have suggested this relationship as follows.

$$\mathrm{Ln\,N(M)} = \alpha - \beta \mathrm{M} - \gamma/\mathrm{M}$$

where N and a represent the annual event rate and β and b represent the seismicity of the study area. Using the coefficients, A and B obtained in the above relations, it is possible to determine the seismic activity of each region and thus the regions with similar seismicity are located in a seismic province.

6.2.1.1 Seismic Hazards and Related Factors in Dezful Region

Based on the information extracted from the earthquake bulletins that have been extracted in a radius of 50 km of this region, the seismicity of the region was studied according to the Gutenberg–Richter method (Fig. 6.7). Considering the driven fault that can cause destructive earthquakes, the following parameters can be presented:

Fault length: 50 km, rupture length: 20 km, maximum earthquake based on fault activity: (MCE) 6.7 degrees Richter, maximum horizontal acceleration g: (PGA) 0.43, maximum horizontal acceleration in the design range: 0.08 g 0 (considering the fault activity increases).

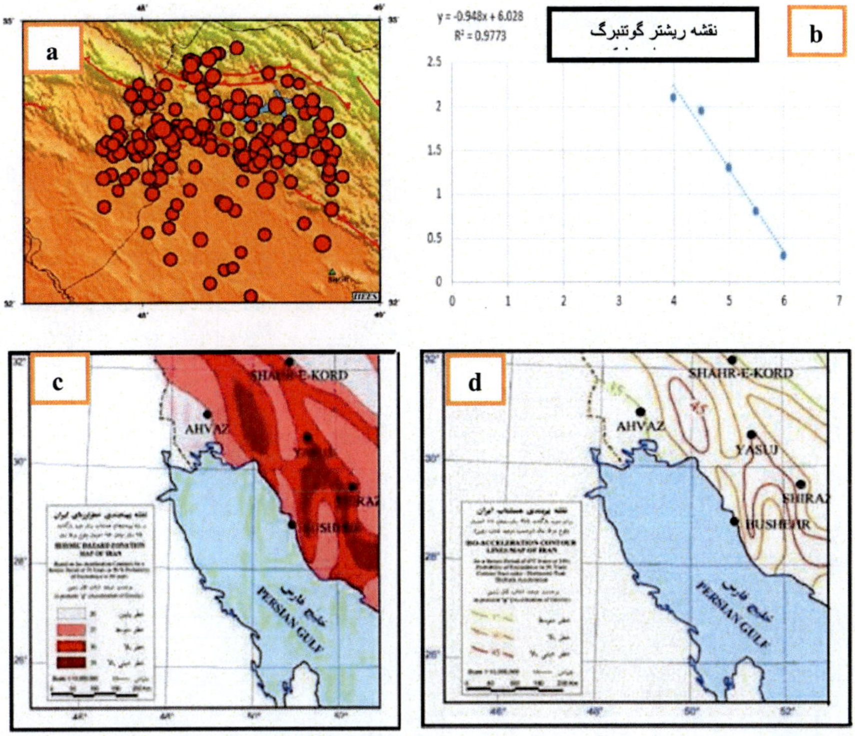

Fig. 6.7 Seismicity studies of Dezful region with the help of data from the Seismological Center of the country (Abbasi et al. 2015). **a** Position of seismic centers at a distance of 50 km in Andimeshk region, **b** Gutenberg Richter diagram for the study area, **c** Seismic hazard zoning map in the study area, **d** Seismic acceleration line zoning in the study area

6.2.1.2 Seismicity Studies and Related Factors in Ramhormoz Region

Ramhormoz and surrounding areas Due to its proximity to faults such as mountain forehead, Ramhormoz fault has always been exposed to earthquakes. Due to the presence of Gachsaran Formation and uncoordinated deformation in Dezful Formation, the surface epicenter of many earthquakes does not match the surface effect of the fault and most of the active seismic faults are latent. Also, the flat-ramp geometry of the faults has caused the seismic part of the faults to be significantly different from their surface effect in terms of location. For example, the separation level of Ramhormoz fault, despite its considerable length, is located within the Gachsaran Formation, while seismic faults have affected Asmari and older hard structures. Table 6.7 and Fig. 6.8 show the statistics and location of earthquakes larger than 5 Richter within a radius of 50 km of Ramhormoz city. Based on the information extracted from the seismic bulletins that have been extracted in a radius of 30 km in this area, the seismicity of the area was studied according to the Richter Gutenberg method, the results of which are shown in the chart and Table 6.7.

Table 6.7 Earthquake data located in a radius of 50 km of Ramhormoz city with the help of data from the Seismological Center of the country (Abbasi et al. 2015)

No	Date (yyyy/mm/dd)	Time (UTC)	Latitude	Longitude	Depth	Magnitude	Reference
1	1948/07/30	03:30:14.0	31.41	49.12	43	Ms: 5.5	NOW
2	1964/08/12	19:26:25.0	30.86	49.66	15	mb: 5.2	EHB
3	1976/09/05	16:43:14.0	31.40	49.96	20	mb: 5.1	EHB
4	1985/03/27	02:06:44.0	31.60	49.92	50	mb: 5.2	EHB
5	1985/09/18	00:10:33.0	31.58	49.47	15	Mw: 5.3	EHB
6	1995/04/22	00:21:51.0	30.97	49.92	30	Mw: 5.3	EHB

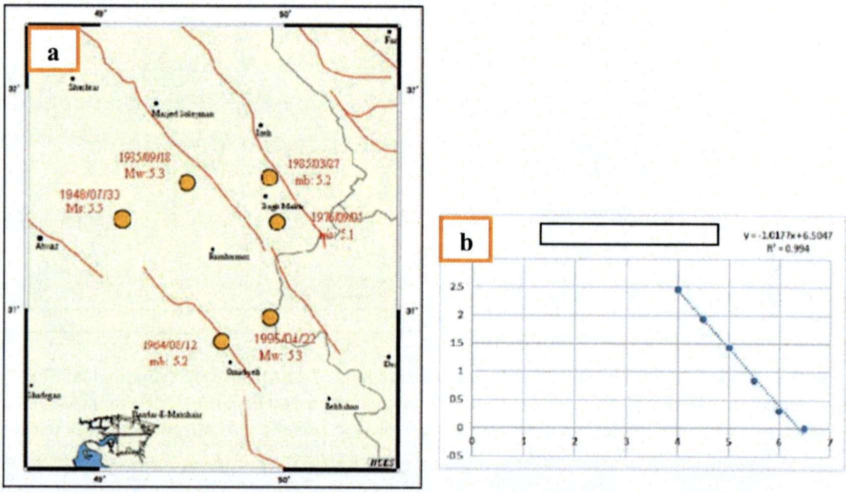

Fig. 6.8 Seismicity studies of Ramhormoz region (prepared with the help of data from the International Institute of Seismology and Earthquake Engineering Abbasi et al. 2015) Ramhormoz, b Richter Gutenberg for the study area in southwestern Iran

6.2.1.3 Earthquake Parameters

Considering the Ramhormoz driven fault, the following parameters can be presented:

Fault length: 70 km, rupture length: 22 km, maximum earthquake based on fault activity: (MCE) 6.2 degrees Richter, maximum horizontal acceleration g: (PGA) 0.43, maximum horizontal acceleration in the design range: 0.08 g (considering the activity of Ramhormoz fault).

In addition to seismicity in Ramhormoz study area, the mass movements of the area are mostly seen in the form of rock fall and landslide. Falling of rocks has occurred mostly in the steep slope of Gachsaran Formation so that these slopes are covered with sunken blocks. This phenomenon can be seen on both sides of the Ala River and other places where the Gachsaran Formation forms high precipices

Fig. 6.9 Landslide occurrence in Gachsaran Formation along the Ala River, in southwestern Iran

along the waterways. Landslides are mostly observed in unstable slopes of Gachsaran Formation, such as Ramhormoz drift front (Fig. 6.9).

References

Abd El-Aal AEAK, Hagag W, Sakr K, Saleh M (2020) Seismicity, seismotectonics and neotectonics in Egypt. In: Hamimi Z, El-Barkooky A, Martínez Frías J, Fritz H, Abd El-Rahman Y (eds) The geology of Egypt. Regional Geology Reviews, Springer, Berlin

Asare-Donkor NK, Adimado AA (2020) Groundwater quality assessment in the northern and upper east regions of Ghana. Environ Earth Sci 79:205

Cellone F, Carol E, Pugliese I (2020) Nitrate pollution in dairy farms and its impact on groundwater quality in a sector of the Pampas plain Argentina. Environ Earth Sci 79:258

Ding Z, Koriem MA, Ibrahim SM (2020) Seawater intrusion impacts on groundwater and soil quality in the northern part of the Nile Delta Egypt. Environ Earth Sci 79:313–329

Esmeray E, Gokcekli C (2020) Assessment of groundwater quality for drinking and irrigation purposes in Karabuk province, Turkey. Environ Earth Sci 79:325–342

Gharahi N, Zamani R (2020) Evaluation of groundwater quality for drinking purposes: a case study from the Beheshtabad Basin, Chaharmahal and Bakhtiari Province Iran. Environ Earth Sci 79:82

Guevara-Mansilla O, Lopez-Loera H, Ramos-Leal JA (2020) Characterization of a fractured aquifer through potential geophysics and physicochemical parameters of groundwater samples. Environ Earth Sci 79:352–367

Hakanson L (1980) Ecological risk index for aquatic pollution control, a sedimentological approach. Water Res 14:975–1001

Jung HB (2020) Geochemical and hydrological study of coastal groundwater discharging to an urban estuary in northern New Jersey. Environ Earth Sci 79:158–169

Lin H, Li H, Yang X (2020) Comprehensive investigation and assessment of nutrient and heavy metal contamination in the surface water in China. J Ocean Univ 19:843–852

Liu W, Zhao J, Ouyang Z, Soderlund L, Liu G (2016) Impacts of sewage irrigation on heavy metal distribution and contamination in Beijing China. Environ Intl 31(6):805–812

Mehdizadeh R (2019) Kinematic evaluation of the Kazerum fault system within the zagros fold-and-thrust belt, Iran. Geotectonics 53:774–785

Muller G (1969) Index of geoaccumulation in the sediments of the Rhine River. GeoJournal 2:108–118

Park S, Choi M, Jang D (2020) Distribution and sources of dissolved and particulate heavy metals (Mn Co, Ni, Cu, Zn, Cd, Pb) in Masan Bay Korea. Ocean Sci J 55:49–67

Pourmorad S, Harami R, Solgi A, Ale Ali M (2021) Sedimentological geochemical and hydro-geochemichal studies of alluvial fans for mineral environment purposes. Lithol Mineral Res 56:89–112

Wang Q, Dong S, Wang H (2020) Hydrogeochemical processes and groundwater quality assessment for different aquifers in the Caojiatan coal mine of Ordos Basin, northwestern China. Environ Earth Sci 79:199–219

Chapter 7
Remote Sensing Studies

Abstract Since the field of door measurement may be different from the expertise of many readers of this book, this chapter first introduces the basic definitions of this science and finally some applications of this field in the study of alluvial fans, especially for mineral applications and Environmental paid.

Keywords Electromagnetic spectrum · AHP · Spectroscopy · Spectrum of rocks

7.1 Primary Studies

Remote Sensing Remote Sensing can be called ground-based information acquisition and imaging technology using aeronautical equipment such as airplanes, balloons or space equipment such as satellites (Jensen 2015). Remote sensing is the science of extracting information from terrestrial objects indirectly by means of a sensor (Clark and Roush 1984). The output is initially an image by each sensor (Gomeh et al. 2014). Each satellite usually has several sensors that do the imaging. The altitude of remote sensing satellites is often more than 700 km and less than 1500–1600 km. Which is less than the height of GPS satellites; because less positional accuracy is required? Each sensor has a different application and is designed according to the intended application. For example, a meteorological sensor has lower spatial accuracy (Clark and Roush 1984).

Simply put, remote sensing is the science and art of obtaining information about any subject under study by a tool that is not in physical contact with it (pourmorad et al. 2021a, b). The superior advantage of satellite information over other sources of information is their repetitive coverage of certain areas at regular intervals (Jensen 2015). In remote sensing, the transmission of information is done using electromagnetic radiation (EMR). The figure schematically shows the general process and the elements that affect the electromagnetic remote sensing of earth resources (Clark and Roush 1984). The two basic processes include data acquisition and analysis (Fig. 7.1).

The analyzer (user) with the help of data collected by the sensor extracts information about the type, amount, location and condition of various land resources, then this information (in the form of maps, printed tables or computer files) with other

S. Pourmorad and A. Mohanty, *Alluvial Fans in Southern Iran*, Advances in Geographical and Environmental Sciences, https://doi.org/10.1007/978-981-19-2045-5_7

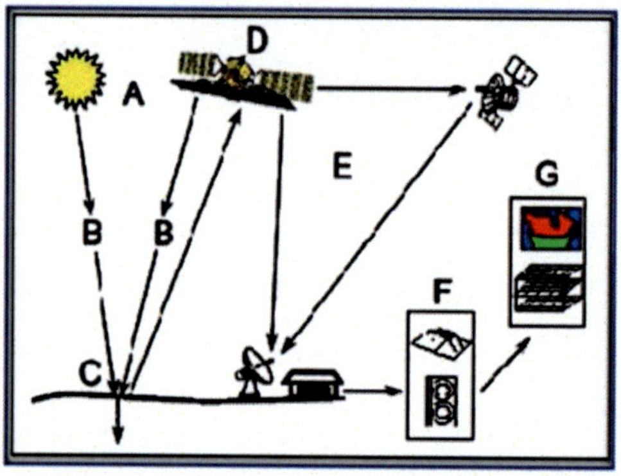

Fig. 7.1 A picture of the factors influencing the formation of satellite images. **a** Source of energy, **b** Emission of energy through the atmosphere, **c** Energy interactions due to collisions with surface terrain, **d** Air or space sensors, **e** Transmission of information obtained, **f** Receive raw data and generate data digitally or visually, **g** Data analysis process

layers of information are integrated into a Geographic Information System (GIS) and prepared for user consumption (Jensen 2015).

7.1.1 Electromagnetic Spectrum

Electromagnetic Radiation is due to atoms and molecules in materials. Atoms contain positively charged nuclei surrounded by orbital electrons that have separate energy levels. The transfer of electrons from one plane to another causes radiation of distinct wavelengths. The result is a spectrum called the electromagnetic spectrum. This spectrum (EMR) reflected from an object is a common source of remote sensing data.

In remote sensing, electromagnetic waves are classified based on the position of their wavelengths in the electromagnetic spectrum. The most common unit used to measure wavelengths in the electromagnetic spectrum is the micrometer. One micrometer is equivalent to one-millionth of a meter. It should also be noted that the segments of the electromagnetic spectrum used in remote sensing are located along a continuous spectrum whose values differ from each other by as much as ten power (sequentially).

Remote sensing technology covers a wide range of electromagnetic spectrums, from very short wavelengths (gamma rays) to very long waves (radio waves). The wavelength range of the electromagnetic spectrum has a range with different names from gamma rays, X-rays, ultraviolet rays, visible light, infrared rays to radio waves (from shorter to longer wavelengths, respectively). The visible part of such a graph is extremely small because the spectral sensitivity of the human eye is between 0.4

and 0.7 μm. Blue is approximately 0.4–0.5 μm, green is approximately 0.5–0.6 μm and red is approximately 0.6–0.7 μm (Jensen 2015).

7.1.1.1 Electromagnetic Spectrum Ranges in Remote Sensing

The range of electromagnetic spectrum is visible to the human eye (signals are transmitted to the brain through eye receptors, and the difference between them gives humans a sense of color recognition.)

Ultraviolet energy is connected to the blue light end of the visible spectrum. At the end of the red light in the visible spectrum, there are three types of infrared waves, which are.

1. Near infrared: from 0.7 to 1.3 μm
2. Middle infrared: from 1.3 to 3 μm
3. Thermal infrared: more than 3 μm.

At longer wavelengths (1 mm–1 m), there is a shortwave (microwave) part of the spectrum.

Most conventional sensing systems operate in one or more parts of the visible, infrared or microwave parts of the electromagnetic spectrum. In other words, each sensor system is sensitive to specific areas of the electromagnetic spectrum and records some of the spectral properties of objects.

For example, conventional cameras are sensitive to the energy of visible light and close to it, i.e. wavelengths of 0.3–1.2 μm; Thermal infrared scanning sensors are generally sensitive to wavelengths between 1 and 2 μm, and radar devices are sensitive to bands with much longer wavelengths (mm and meters).

– Relationship between wavelength with energy and frequency: shorter wavelength, higher energy and frequency and vice versa.
– Relationship between wavelength with energy and frequency: Most energy and frequency and short wavelength waves are in the visible range.

7.1.2 *Factors Influencing the Formation of Satellite Imagery*

When radiant energy strikes any object on the earth's surface, energy is reflected, transmitted or absorbed by that substance in three main forms.

The properties of the energies generated vary greatly depending on the conditions and type of material on Earth, which means that some objects have reflective properties against a particular wavelength but absorb and transfer energy at another wavelength. A set of such phenomena creates specific colors or tones on different images and allows the eye to distinguish different shapes in images.

The properties of the energies generated vary greatly depending on the conditions and type of matter on Earth, and these differences lead to the identification of various phenomena by the mass. The electromagnetic wave radiation curve for the items mentioned in the image is a spectral reflection diagram of an object as a function of its wavelength.

Energy interaction in the atmosphere and the earth:

Sunlight is affected, reflected, diffused or absorbed by molecules and suspended particles in the atmosphere as it passes through the atmosphere. This method of changing the intensity of sunlight creates colors. For example, the blue color of the sky during the day is due to the emission of blue spectra in the atmosphere.

The reason for the blue of the sky on the day is due to the blue spectrum propagation in the atmosphere (all shorter wavelengths are propagated over a distance and only longer wavelengths reach the Earth's surface).

All matter is made up of atoms and molecules with a specific composition. Thus, each material absorbs, reflects, or emits electromagnetic radiation in a single form and under a specific wavelength that is related to its internal energy level, which is called the unit properties of the material or the spectral properties. For example, the reason some leaves appear to be green is that chlorophyll absorbs the blue and red spectra and reflects the green spectrum. The green color of vegetation is the greatest reflection of the green spectrum (Jensen 2015).

7.1.3 Platforms

The carrier of telemetry sensors is called a platform. Typical types are satellites and airplanes, but other devices such as radio-controlled aircraft and balloons are also used for remote sensing at low altitudes. In general, the key factor for choosing a platform is the height because the ground resolution is determined and it also depends on the instantaneous field of view (IFOV) of the sensor on the platform (Rangzan et al. 2018).

In the study of alluvial fans, according to the main purpose of the studies, various data and methods of remote sensing can be used, such as thermal remote sensing studies, radar remote sensing studies, etc., which explain all methods and methods beyond the power of this It is a book. Therefore, in this chapter, we have tried to use two methods of remote sensing, which while being simple, have many applications in these studies. These two methods include the Analytic Hierarchy Process (AHP) method and the spectroscopy method, each of which is practically described in the following section.

7.2 Analytic Hierarchy Process (AHP)

According to Forman and Saul (2001), the analytic hierarchy process (AHP), also analytical hierarchy process, is a structured technique for organizing and analyzing complex decisions, based on mathematics and psychology.

According to Salavati et al. (2016), in this technique with Individual experts, experiences are utilized to estimate the relative magnitudes of factors through pairwise comparisons. In fact, each of the respondents compares the relative importance of each pair of items using a specially designed questionnaire. The AHP method understands that although there are several criteria for our questions, but the magnitude of each criterion may not be equal. Hence, while evaluating alternative solutions, weights need to be attached to the criteria to ensure reaching the correct conclusion. This may seem obvious.

According to Pourmorad et al. (2021a, b), this technique can be used in various disciplines and specialties such as petroleum engineering, geology, ecology, management and other sciences that require the analysis of several complex factors.

According to the authors of this book, in this technique, when an important question about an important phenomenon is considered, we should try to identify the various factors that affect it first, and then for each of these factors, the importance and impact on Define on the final result we expect. For example, when we decide to identify the karsts of an area that are important sources of groundwater, we must first identify the most important factors influencing the formation of karsts (e.g. lithology, distance from the fault, distance from the waterway and the slope of the area). Then we have to validate each of these factors based on their importance and role in karst formation (for example, carbonate lithology takes the most weight).

Based on the weight of each of the mentioned factors, the final map of the karstification potential of different parts of the study area is determined. In the following sections, this technique is better known by mentioning the practical example that we have done in the southwest of Iran regarding the Flooding of one of the largest and most important provinces of Iran (Khuzestan province).

7.2.1 Steps of Analytic Hierarchy Process (AHP)

According to the experience and expertise of the authors of this book, there are four main steps to get acquainted with how this software works, which we will review in the following section:

1. Creating a hierarchical diagram:

According to the authors of this book in this step, first the research factors should be extracted from different sources or experts should be asked. After extracting the factors and options, he divided the problem into standard levels and if there are subcriteria and options. The existence of a criterion in the AHP model is essential, i.e. a hierarchical model will not be created without a criterion.

2. Formation of even comparisons matrix:

According to the experience and expertise of the authors of this book, in this stage, the elements of each level are compared in pairs at a higher level than other related elements, and even comparisons matrices are formed.

3. Compatibility rate calculation:

The incompatibility rate indicates whether the comparisons are stable or not. In software specific to the AHP method, this rate is automatically calculated by the software. If this rate is less than 0.1, it indicates matrix compatibility, and if it is higher than 0.1, pairwise comparisons should be revised.

When a large number of factors are involved in a decision-making problem, the inconsistency rate is very high and eliminating it requires many changes in the pairwise comparison matrix. It is recommended to use the improved AHP method in such cases.

4. Weighting by AHP method:

In order to perform the AHP method, we must first weigh the factors influencing this research in order to determine the importance of the research criteria and options. In this step, using different weighting methods, we get the relative weight of the criteria. According to Salavati et al. (2016), there are different methods for weighting, including line sum method, column sum, arithmetic mean, geometric mean, special vector method and square sum method.

In a model that includes criteria and options, we must first obtain the weight of the criteria relative to the goal, and then the weight of the options relative to each of the criteria. Finally, by multiplying the criteria weight by the relative weight of the options, the final weight of the options is obtained.

All steps of the AHP method are performed automatically by two software, expert choice or super decision, and there is no need for any manual calculation by the researcher. Since the step-by-step instruction of this software is beyond the scope of this book, the authors of this book recommend that researchers and students learn this software from reputable sources.

The following section provides a practical example of the AHP method for flooding in alluvial fan-related areas in southwestern Iran.

7.2.2 Flood Studies with the Help of Remote Sensing Data

One of the most important risk factors in alluvial fans is flood hazards, especially in the vicinity of alluvial fans in urban areas (Costa 1983).

In this section, the aim is to find a model in this type of study in southwestern Iran with the help of effective information layers on flooding such as slope, rainfall intensity, rainfall, waterway density, lithological information, fault density flooding map of the studied areas should be prepared and examined.

In this study method, after examining each of the mentioned information layers in the study areas and preparing the output of each information layer, the mentioned factors are weighed and then by the method of fuzzy hierarchical analysis process to investigate flooding and prepare a flood map. Rise has been addressed in these areas and has finally been weighed against field evidence to ensure the data obtained.

7.2.2.1 Slope Measurement in Dezful and Ramhormoz Areas

Basin slope is one of the most important factors that controls the surface flow time and water concentration in the river or on other sloping surfaces such as alluvial fan deposits (Nagender Nath 2020). The slope of a basin affects the amount of infiltration, surface flow rate, soil moisture and the amount of groundwater in the basin (Casalbore et al. 2020). Basically, steeply sloping basins often have flood flows (Nagender-Nath 2020). Due to the high slope, the speed of water flow increases and as a result, the rate of infiltration and evaporation decreases, after a short period of rain, it turns into flow and flood occurs. The hydrograph range of such basins is short and their peaks are high (Askandari Nejad et al. 2015). For this reason, in mountainous areas, if the permeability is low, water flooding will be high and flood damage will be high (Omidvar et al. 2010).

In this study, after preparing the digital elevation model (DEM), using the Slope function, the slope map of Khuzestan province was prepared with the cooperation of GIS and tectonics experts of the Geology and Mineral Exploration Organization of Ahvaz Center for the whole of Khuzestan province (Fig. 7.2a, b).

The results show a greater slope of Dezful study area than the Ramhormoz area. This slope, which is quite evident in field studies, is a high-risk factor in creating surface runoff from the upstream parts of the alluvial fan to the downstream parts of the alluvial fan (urban areas of Andimeshk and Dezful), which except for the impact Direct transport of sediments will cause many hazards in these areas.

Of course, what reduces the risk of flooding in Andimeshk region to some extent is good sorting and consequently good permeability of particles in this region, especially in the upper part of them, which is the permeability in the part that has the

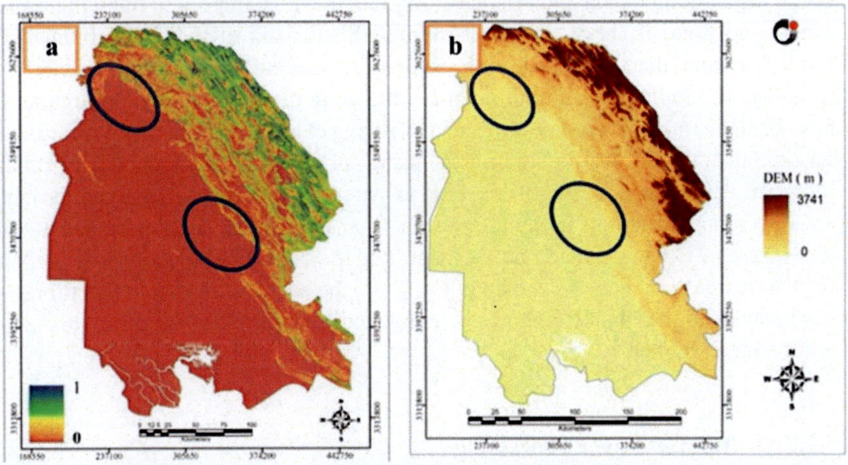

Fig. 7.2 Slope layer processing in Dezful and Ramhormoz areas. **a** Slopemap of Khuzestan province with separation of study areas, **b** Digital elevation model (DEM) map with separation of study areas (study areas of Dezful and Ramhormoz are marked in the picture)

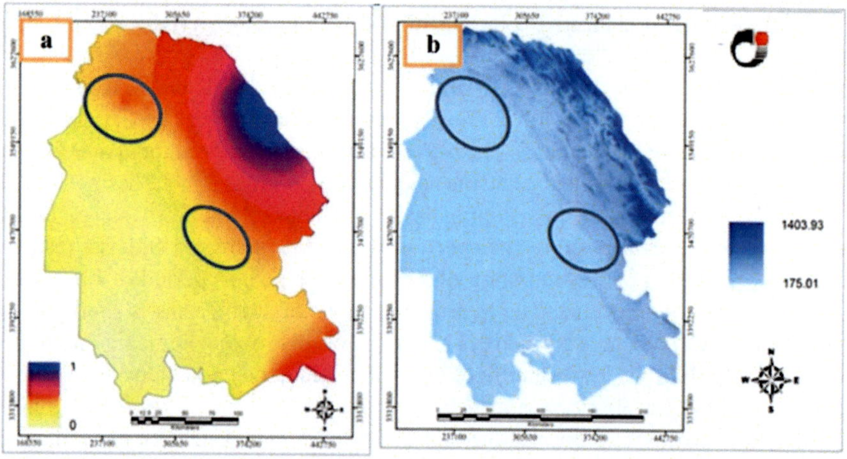

Fig. 7.3 Processing of rainfall intensity layers and rainfall. **a** Map of average annual rainfall intensity of Khuzestan province and study areas, **b** Map of average annual rainfall of Khuzestan province and study areas (study areas of Dezful and Ramhormoz are shown in the picture)

highest slope of these alluvial fans. It has (direct alluvial fan) has a direct effect and reduces runoff and flood flows caused by them.

7.2.2.2 Investigation of Rainfall Intensity

Rainfall and hail are known as the primary source of most floods (Sime and Demissie 2020). But among the characteristics of rainfall, the most important thing that causes runoff and flood is the intensity of rainfall (Shemshaki et al. 2011). Intensity of rainfall greater than the infiltration capacity of the catchment causes runoff and flooding. The higher the rainfall intensity, the more runoff is created throughout the basin and the more severe flooding occurs (Zhang et al. 2020). To obtain information on rainfall intensity in the study area, using the average maximum daily rainfall intensity, the average period of 25 years in synoptic stations of Khuzestan province and also adjacent stations in Ilam, Lorestan, Chaharmahal and Bakhtiari, Kohgiluyeh provinces. In Boyer Ahmad and Bushehr, a map of rainfall intensity was prepared for Khuzestan province. As shown in Fig. 7.3a, the average rainfall in Dezful region is somewhat higher than Ramhormoz region and as a result flood risks in this region will be much higher.

7.2.2.3 Rainfall Layer Processing

In addition to rainfall intensity, rainfall is another influential meteorological feature in the flood potential of an area (Sime and Demissie 2020). Flood is directly related

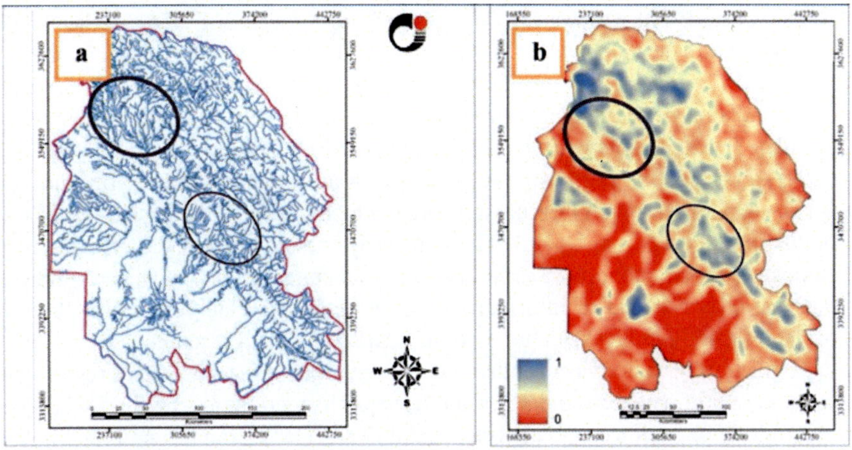

Fig. 7.4 Processing the density of waterways in Khuzestan province and the location of the studied areas, **a** Location map of waterways, **b** Density map of waterways in Khuzestan province in southwestern Iran (Dezful and Ramhormozstudy areas are marked in the picture)

to rainfall (Zhang 2020). In general, the more rainfall one region has than the other, the more floods will occur there (Alizadeh 2020). In order to determine the effect of rainfall on flooding zonation, the average annual rainfall of meteorological stations in Khuzestan province was used. Due to the direct relationship between ground level and rainfall, and on the other hand due to the lack of sufficient stations in the region, the altitude-rainfall equation of Khuzestan province was obtained using the average annual rainfall of stations according to the following equation (Alizadeh 2020):

$$P = 0.3285\,h + 175.01$$

Then, using the equation obtained and the digital elevation layer (DEM) of the region, a map of the average annual rainfall of the province was prepared and a comparative information was obtained between the rainfall in Dezful and Ramhormoz (Fig. 7.3b). According to the obtained data, the average annual rainfall in Ramhormoz region is 220 ml and in Dezful region is 340 ml.

7.2.2.4 Calculation of Waterway Density in the Studied Areas

The network of basin waterways shows how runoff is discharged from the basin surface (Malkian et al. 2012). The lower the surface permeability of the basin, the greater and more developed the waterway network of the basin (Liu et al. 2017). Simply put, a network of waterways is a reflection of the volume of runoff and floods in an area. Therefore, the more the waterways of a basin evolve and develop, the more flooded that basin is. In order to prepare this map, first, the digital file (Shapfile) of the province's waterways, prepared by the Khuzestan Water and Electricity Organization, was combined with the waterways obtained from the DEM layer in the ArcGIS environment. The output layer in Google Earth software was then validated with real streams, and new streams were added and some removed (Fig. 7.4a). From the calculation of the length of the final lines, the density layer of the canals was calculated using the Density function in the ArcGIS environment (Fig. 7.4b).

As can be seen in Fig. 7.4b, the density of waterways in Ramhormoz study area is somewhat higher than Dezful study area, which indicates the possibility of more runoff in this area than Dezful. The conclusion obtained from the changes in the slope of the basin mentioned earlier is somewhat contradictory and clearly indicates that one or two layers of information cannot be an accurate conclusion about the basin, especially in areas dealing with detrital sediments. Presented.

Extraction of waterways in the studied areas shows that in Dezful region these waterways originate from Bakhtiari and Aghajari formations and Lahbari section and in Ramhormoz region originate from Bakhtiari and Gachsaran formations (Fig. 7.4).

7.2.2.5 Petrological Information Processing of the Studied Areas

One of the effective factors in creating runoff in the studied basins is the resistance of the formations forming the surface of the basin against erosion and water permeability, in which each formation shows a different resistance (Feng 2019). The strength of formations depends on the type and composition of the constituent minerals, the way of crystallization, the amount of pores, stratification, the way of outgrowth and erosion, interlayer, etc. (Xi et al. 2017). To prepare this layer and classify different lithologies of the basin in terms of flooding potential, the geological map of Khuzestan province was used and the results of processing this information layer in Fig. 7.5 is visible.

Studies conducted in Ramhormoz and Dezful regions show that in Ramhormoz region due to the existence of erosive formations such as Gachsaran with evaporitic lithology (gypsum and anhydrite) at the top of Ramhormoz alluvial fan and due to the high solubility of this type of lithology This region is more likely to cause runoff than Dezful region, which includes Aghajari and Bakhtiari Formations with sandstone lithology. These results are consistent with the results of canal density studies.

7.2.2.6 Compaction Layer Processing of Faults

The most important phenomenon in creating runoff and flooding is permeability (Qanawati et al. 2011). Structural geology is another factor that causes changes in the permeability of rock units (Shoorangiz 2020). To the extent that it sometimes plays a more influential role than lithology. The tectonic factor causes faulting and fracture in different rock units by applying stress (Hagos 2020). These fractures create secondary porosity in the rock, which increases permeability and ultimately

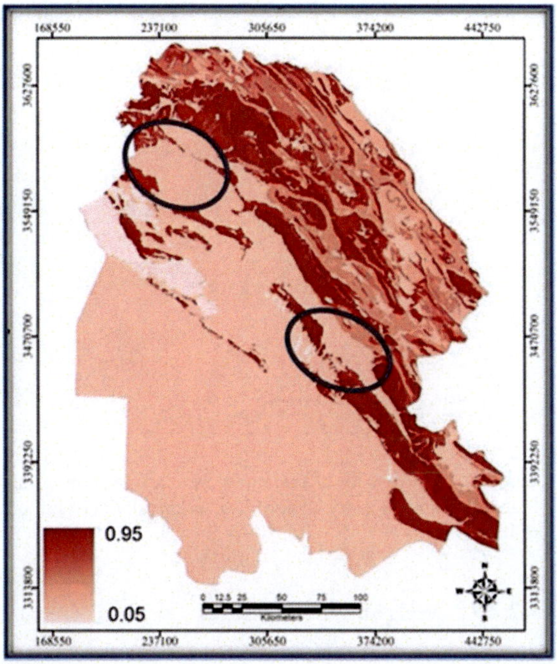

Fig. 7.5 Petrological information processing of the studied areas, the standard map of the geological layer of the studied area (the studied areas of Dezful and Ramhormoz are marked in the picture)

reduces flooding. Of course, the higher the density of these fractures, the greater the permeability (Shoorangiz 2020).

Since geological hazard studies in the study areas were conducted in the form of general studies of Khuzestan province and due to the size of the study area, it is practically impossible to measure fractures and determine the location of the fracture. But since the presence of fractures depends on the presence of faults and thrusts, on a large scale by locating the faults and calculating their density, the highest fracture density can be logically achieved. To prepare this layer, the faults drawn in the geological maps of the province were merged with the faults drawn using Google Earth software and the distribution map of the faults of the province was prepared (Fig. 7.6).

Finally, using the fault layer and applying the Density function in ArcGIS environment, the fault density map of Khuzestan province was prepared and finally the study areas were comparable with each other. With the prepared map, the amount of corrosion faults in the studied areas is quite clear. According to these studies, the fault density in Dezful region is higher than Ramhormoz region (Fig. 7.6b). According to these studies, the higher density of faults in Dezful region has increased the permeability in this region and as a result reduces the risks of runoff in alluvial fans of this region. This conclusion was previously discussed in the section on slope studies and is fully consistent with this conclusion.

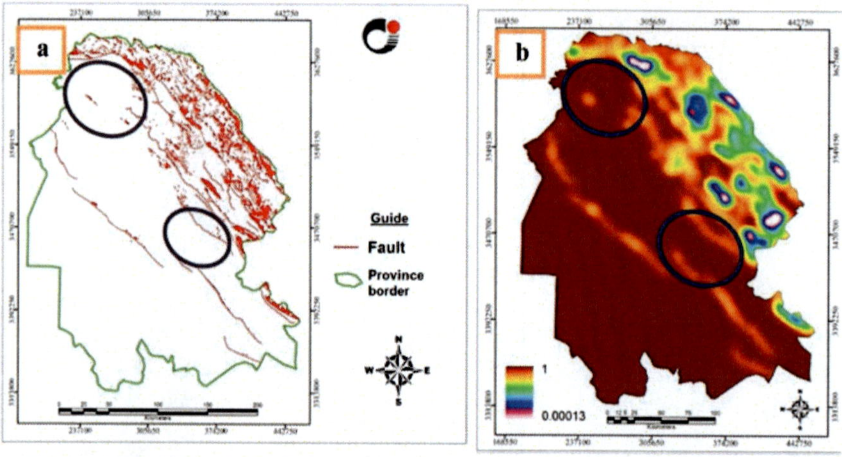

Fig. 7.6 Fault density of the study area. **a** Distribution map of faults of Khuzestan province, **b** Standard map of density layer of Khuzestan province fault in southwestern Iran, Geological Survey of Iran

7.2.2.7 Weighing the Criteria

The weight of each factor indicates its importance and value compared to other factors in flood zoning potential zoning operations (Ghodsipour 2006). In fact, conscious and correct selection of weights is a great help in the direction of proper zoning (Ghazanfari and Rezaei 2006).

After applying the membership functions to the layers, weighting of the layers becomes necessary due to the fact that each of them has a different effect on the flood potential. Invoicing operations are performed using expert knowledge. In this method, taking into account the opinions of experts in the field of application and considering the characteristics of the study area, the appropriate weight for each factor has been obtained. To calculate the weight of the criteria, two methods of Analytic Hierarchy Process (AHP) and Fuzzy Analytic Hierarchy Process (FAHP), which are knowledge-based methods, have been used. For this purpose, fuzzy numbers and fuzzy scales are used first. The following table shows the linguistic significance and fuzzy numbers in the form of triangular fuzzy scales (Table 7.1).

The validity of pairwise comparisons of the main criteria, sub-criteria of hydro-climate, sub-criteria of topography and surface cover and sub-criteria of geology was performed by obtaining the incompatibility coefficient less than 0.1 and the values of the incompatibility coefficient being 0.07, 0.02, respectively, 0 and 0 are obtained.

After forming the pairwise comparison matrix and completing it with the mentioned methods, the weight of each criterion is determined. To do this, use the program written in the MATLAB software environment and Expert Choice software, and by entering the data of the pair comparison table that are fuzzy, the weight of each criterion is specified (Tables 7.2, 7.3, 7.4 and 7.5).

Table 7.1 Triangular fuzzy numbers used in the study areas

Linguistic significance	Non-fuzzy number	Triangular fuzzy scale	Inverse non-fuzzy number	Inverse fuzzy scale
Exactly equal importance	1	1, 1, 1	1	1,1,1
Almost equal importance	2	1.2, 1, 3.2	0/5	0.66, 1, 2
A little more important	3	1, 3.2, 2	0/33	0.5, 0.66, 1
More important	4	3.2, 2, 5.2	0/25	0.4, 0.5, 0.66
Much more important	5	2, 5.2, 3	0/2	0.33, 0.4, 0.5
Much more important	6	5.2, 3, 7.2	0/17	0.29, 0.33, 0.4

Table 7.2 Pair comparison and weight of main criteria

FAHP	AHP	Geology	Hydroclimate	Topography and surface coverage	
0/2934	0/268	3	0/33	1	Topography and surface coverage
0/5304	0/614	4	1	3	Hydroclimate
0/1762	0/117	1	0/25	0/33	Geology

Table 7.3 Binary comparison and weight of hydroclimatic standard sub-criteria

FAHP	AHP	Rainfall intensity	Waterway density	Rainfall	
0/284	0/333	3	0/33	1	Precipitation
0/1906	0/140	4	1	3	Waterway density
0/5254	0/528	1	0/25	0/33	Rainfall intensity

Table 7.4 Binary comparison and weight of sub-criteria of topography and surface coverage

FAHP	AHP	Land use	Slope	
0/7139	0/8	4	1	Slope
0/2861	0/2	1	0/25	Land use

Table 7.5 Binary comparison and weight of geological criteria sub-criteria

FAHP	AHP	Fault density	Lithology	
0/7139	0/8	4	1	Lithology
0/2861	0/2	1	0/25	Fault density

7.2.2.8 Overlap

Since the effective criteria and sub-criteria have different weights and all of them must participate in the overlap, the index overlap method or the Weighted Linear Combination method according to the following formula has been used for this task (Burrough 1990).

$$A_i = \sum w_j x_{ij}, \sum w_j = 1$$

In this method, the standardized layer obtained from each criterion is multiplied by the weight of that criterion. The analyst or decision maker directly weighs the criteria based on the relative importance of each criterion under consideration. Then, by multiplying the relative weight by the value of that attribute, a final value for each option is obtained. Once the final value of each option has been determined, the options with the highest value will be the most appropriate option for the target (Burrough 1990).

This is done for all criteria and sub-criteria, and the new layers obtained are overlapped using various functions such as AND, OR, SUM, Product and GAMA. After overlapping the main criteria with different functions, finally the output of a function is considered as the best option, according to field and statistical evidence, for zoning of flood potential. At this stage, the output of FAHP method is considered as the final output (Figs. 7.7 and 7.8).

7.2.2.9 Map Validation with Field Evidence

In order to interpret, analyze and validate the prepared zoning, different areas of the province that have statistics and evidence of floods have been studied on the final zoning map.

In Andimeshk region, due to the good cementation of Bakhtiari conglomerate (calcareous cement), this formation has created impermeable surfaces and rocky and steep heights that have a high flood potential. The result of this high potential is the occurrence of severe floods and the emergence of deep and wide waterways in the region, which is a proof of the accuracy of the danger of this area (Fig. 7.9a, b).

In the northeast of Ramhormoz city, there are heights of Gachsaran and Bakhtiari formations. Occurrence of frequent floods, active waterways with high sediment load, along with the construction of flood distribution projects and earth dams indicate the high potential of floods in the region and the accuracy of the prepared map. Bakhtiari and Aghajari formations are among the high-risk areas due to sandstone lithology and Gachsaran Formation with evaporitic nature is in low to medium risk category. In Ramhormoz region, flooding has been observed in different seasons and has caused a lot of damage, which confirms the results obtained (Fig. 7.10).

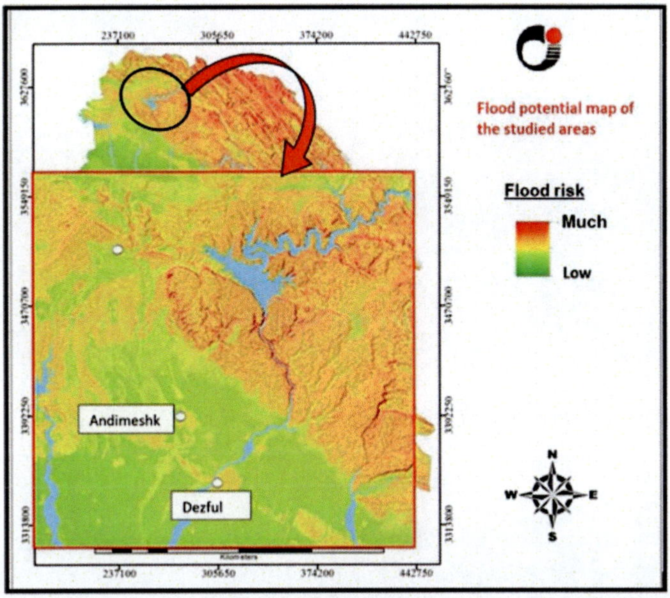

Fig. 7.7 Flood potential potential zoning by Fuzzy Analytic Hierarchy Process (FAHP) method in Andimeshk study area, southwest of Iran

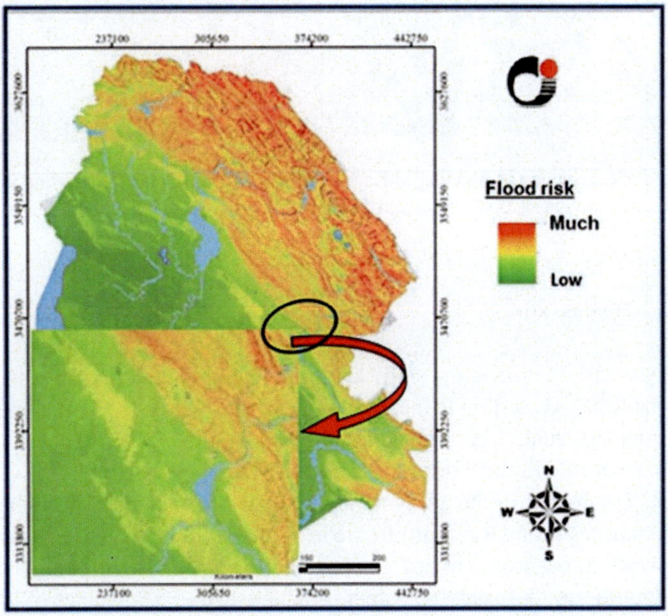

Fig. 7.8 Flood potential zoning by Fuzzy Analytic Hierarchy Process (FAHP) method of Ramhormoz study area, southwest of Iran

Fig. 7.9 Field evidence of flood control in the study area in southwestern Iran

Fig. 7.10 Field evidence of flooding in Ramhormoz area. **a** Evidence of flooding in Ramhormoz region, **b** Construction of flood dams upstream of Ramhormoz large alluvial fan in southeastern Iran

7.2.2.10 Discussion

The study of risk factors in the study areas provides significant results in these areas for the first time. According to these studies, various factors such as slope, lithology, rainfall intensity, runoff, earthquakes and other factors affecting the area can play a very important role in the risk of these alluvial fans in the area.

Studies conducted in the study areas in terms of seismicity show that the study areas of Andimeshk and Ramhormoz are in relatively low seismic hazard ranges. The earthquakes that occurred around the study areas, especially in Ramhormoz region, have magnitudes of 5–6 and 3–5° on the Richter scale. The distance of the epicenter of these earthquakes from the study area in the mentioned areas varies between 10 and 20 km. Considering the very gentle slope and mostly angular gravel particles in the Ramhormoz large alluvial fan, seismicity in this region has less effect in terms of excitation and particle displacement in the alluvial fans of this region than the alluvial fans in Andimeshk region. It has a large number of alluvial fans with steep slopes and particles with relatively good roundness on the mountain front.

The results of flooding in the study areas by different layers of information show that both areas have a relatively high flood potential that this flood potential in these areas due to the proximity of Andimeshk, Dezful and Ramhormoz cities. The alluvial fans under study can pose very serious dangers to the inhabitants of these cities.

According to studies, the slope of alluvial fans in Dezful region is higher than Ramhormoz region, which has a greater impact on the transport of sediments, runoff, mudflows and other hazardous factors in Andimeshk region. In addition, this increase in slope along with more rainfall and rainfall intensity in Dezful region will increase surface runoff in Dezful region. However, as mentioned above, the high permeability of the upstream part of the Andimeshk alluvial fans has reduced the slope effect in creating surface runoff (due to faster water penetration into the lower parts).

It should be noted that the higher density of waterways in Ramhormoz region than Dezful region can be a good alternative for less rainfall and intensity of rainfall in this region and as a result causes the amount of surface runoff in this region is also significant.

Other effective factors in seismic hazards as well as runoff in the two areas under study are fault densities. According to the studies, the density of faults in Dezful region is higher than Ramhormoz region. These studies increase the density of faults in Dezful region and increase the permeability in this region and thus reduce the risks of runoff. The water flows in the alluvial fans of this region.

From the total information obtained, it can be inferred that both Ramhormoz and Dezful regions have almost the same flood risk because in Dezful region, despite the greater impact of factors such as rainfall, rainfall intensity, slope and fault density. Due to lithology with good permeability, we see a decrease in the volume of surface runoff.

7.2.3 Spectroscopy

The study of the interaction of electromagnetic radiation with a spectroscopic material is called spectroscopy (Piao et al. 2020). Electromagnetic radiation, which is characterized by components such as wavelength, frequency and velocity, has a dual property of particle and wave. In fact, electromagnetic radiation can be thought of as separate particles called photons whose energy is proportional to the frequency of the radiation (Forzieri et al. 2020).

The electromagnetic spectrum ranges from radio waves to gamma rays, which are called different letters depending on their frequency. In ultraviolet/visible spectroscopy, the absorption of electromagnetic radiation excites the electrons in the capacitance layer of atoms and molecules (Sánchez and Navarro 2017). These electron transitions in atoms cause a number of narrow absorption lines, but in molecules they give continuous and wide absorption peaks (Piao et al. 2020).

Spectroscopy is the science that studies the interaction of electromagnetic radiation with matter. In such interactions, electromagnetic radiation can be thought of as a set of separate energy packets called photons. Considered (Forzieri et al. 2020). In addition to particle properties, electromagnetic radiation also has wave properties. This dual property of electromagnetic radiation as a particle and a wave is not only not contradictory but also complementary (Sánchez and Navarro 2017). According to wave theory, electromagnetic radiation is made up of two components, the electric field and the magnetic field. These fields are propagating waves in the environment, perpendicular to each other and also in the direction of the wave propagation (Forzieri et al. 2020).

Quantitative and qualitative studies performed due to the interactions of matter with electromagnetic radiation are called spectrochemical analyzes. In general, these analyzes deal with a wide range of frequencies (Huang et al. 2016). This range ranges from radio waves to gamma rays (Piao et al. 2020). Electromagnetic waves can be read in various letters depending on their frequency: Radiowave, Microwave, Visible, Ultraviolet, UV, X-ray and γ-ray. These letters are arranged in order of increasing frequency (decreasing wavelength).

Spectroscopy was initially started with an intensimeter or Radiometer, which recorded the amount of reflection of the phenomenon as a number only in one wavelength (Forzieri et al. 2020). By changing the wavelength characteristics of this device, we could record each wavelength individually for a phenomenon. With the passage of time and the evolution of devices, spectrometers were created that could generate spectral information in a specific spectral range (Piao et al. 2020). From the combination of a radiometer and a spectrometer, a spectro-radiometer was created that records both intensity and spectral range simultaneously (Forzieri et al. 2020).

7.2.4 Spectral Properties of Mineralogical Components

The following is a summary based on research by various researchers, including Hunt and Salisbury (1971). The discussion is divided into three parts: Visible and near-infrared area, 2. Short infrared, 3. Thermal or thermal infrared.

7.2.4.1 Near Visible and Infrared Region (400–1000 nm)

Spectral properties in this part of the electromagnetic spectrum are created by electronic processes in conductive metals (e.g. Cr, Ni, Cu, Mn, Fe, etc.).

Elements such as Al, Si and various anion groups, such as oxides, hydroxides, carbonates, phosphates, etc., which make up the bulk of the earth's crustal rocks, lack spectral properties in this area.

In this region, iron is the most important component with spectral properties. Ferric ion in Fe–O, a pervasive component in rocks, exhibits strong UV-blue wavelength adsorption due to charge transfer effect (Jensen 2015). This effect leads to a sharp slope falling in reflection to the blue and a general increase in the slope to the infrared.

The maximum wavelength is in the range of 1300–1600 nm (Rangzan et al. 2018).

7.2.4.2 Short Infrared Detection (SWIR)—1000 to 3000 nm

The SWIR region is important in terms of the absorption properties of hydroxyl and carbonates (Jensen 2015). It should be noted that the position and depth of adsorption zones in the SWIR range may change due to crystalline effects. The absorption factors in the SWIR range are given below.

Hydroxyl Ions

Hydroxyl ions are an important component found in stone-forming minerals such as clays, mica and chlorite. The main absorption band of hydroxyl is at 2740–2770 nm and its Everton (harmonizer) is at 1440 nm. The main absorption states 2770 and Everton 1440 interfere with the similar forms seen in the water molecule. However, when hydroxyl ions are combined with aluminum and magnesium, for example, AL–OH and Mg–OH, which are commonly found in hydrated clays and silicates, numerous rapid absorption properties have been observed in the range of 2100–2400 nm.

The vibration band, ALOH is typically located at 2200 nm and the MgOH band at 2300 nm. If the compounds MgOH and ALOH are present together, the maximum absorption usually occurs at 2300 nm and a weaker band at 2200 nm, leading to a binary band; For example, dual absorption bands are observed in kaolinite, but in montmorillonite and muscovite only adsorption occurred at 2300 nm due to MgOH (Gupta 2002).

Also, replacing iron with Al or Mg reduces the step structure in the spectrum.

The strong absorption phenomenon in the range of 2100–2400 nm due to the presence of ALOH and MgOH leads to a sharp decrease in spectral reflection at the wavelength of 1600 nm. If clay is present, this broadband absorption property at 2100–2400 nm is used to detect clay-rich areas (Gomeh 2014).

As the maximum reflectance is about 1600 nm. The ratio of 1550–1750 bands to 2100–2400 nm bands is a strong factor for the identification of clay mineral masses (Jensen 2015). In addition, this band ratio is useful in identifying hydrothermal alteration areas containing clay minerals; For example, kaolinite-muscovite-pyrophyllite-alunite-dikite-montmorillonite-diaspora are in these areas.

Carbonates

Carbonates are commonly formed in the earth's crust in the forms of calcite, magnesite, dolomite, and siderite. Significant carbonate adsorption bands occur in the near infrared at 2350, 1990 and 2550 nm. Adsorption at 1990 nm may interfere with the maximum wave height of molecular water, and adsorption at 2350 nm with a similar property in clay is about 2300 nm. In addition, the presence of siderite is associated with an electronic absorption band due to the presence of iron, which is located in the range of about 1100 nm (Gupta 2002).

Range of Minerals

Minerals are naturally composed of inorganic materials, including compounds of cations and anions. Minerals may have simple or complex chemical compositions; Some ions may be large or small (Gupta 2002). The range of minerals is controlled by the influence of the following factors (Gupta 2002).

- The main spectrum of anions.
- The main spectrum of cations.
- The range of ions that are caused by rare compounds.
- Crystal field effect.

Some examples are given below to clarify the above issue.

Pyroxene

Absorption bands occur due to the presence of iron ions in the near and visible infrared region, and those formed due to the presence of silicate ions are in the thermal infrared range. The position and intensity of the absorption bands are determined by the effect of the crystal field.

Mineral Group of Amphibole, Mica and Clay

Adsorption bands occur either due to iron in the region, VNIR, or due to hydroxyl ions in the SWIR region, or due to silicates in the TIR thermal region. All carbonate or charcoal ores is located in the SWIR and thermal infrared regions. In addition, adsorption due to iron in siderite ($FeCO_3$) iron carbonate is observed in VNIR and due to manganese in rhodochrosite ($MnCO_3$) (manganese carbonate) is also observed in VNIR region (Forzieri et al. 2020).

Clay

Clay has absorption bands in the range of 2300–2200 nm (SWIR) due to the presence of AL–OH/Mg–OH and also shows absorption properties due to the presence of water molecule at 1400 nm. The presence of iron leads to load transfer bands in the nearby sub-red region. In summary, it can be concluded that the spectrum of a mineral is the result of the spectral composition of the main components and the effects of the crystal field.

Spectrum of Rocks

Rocks are aggregates of minerals and are much more complex and different in composition than minerals. It is difficult to define the spectral curves of rocks. However, it is possible. The spectral properties of rocks should be described in detail based on the spectral properties of the main components of minerals (Clark and Roush 1984).

The discussion here is divided into two branches: (a) solar reflection region (b) thermal infrared region.

Solar Reflection Region (VNIR + SWIR)

The spectra of rocks depend on the spectra of the constituent minerals and their properties, such as texture size, density, mixing and so on. Quasi-experimental methods and analytical processes have been proposed for spectral understanding of a mixture of several minerals (pelicans. Mixing can be divided into four types: linear, nonlinear, coating and molecular (Rangzan et al. 2018).

Volcanic Rocks

The graphic granites show the absorption bands at 1900, 1400 and 2200 nm for the absorption bands H2O, OH and hydroxyl compound, respectively. Granites containing black mica (biotite) and granites with less water show weaker OH adsorption bands. Volcanic rocks, which are mostly composed of pyroxene, silicate, amphibole, olivine and mica, contain iron.

In magnetites, amphiboles, and pyroxenes, absorption bands associated with ferric and ferrous ions appear at 700 and 1000 nm. Ultramafic rocks contain large amounts of opaque minerals and minerals containing iron ions, resulting in predominant ferrous bands (Clark and Roush 1984).

Sediment Rocks

All sedimentary rocks usually have water absorption bands at 1400 and 1900 nm. Clay shales have a special absorption property in the range of 2100–2300 nm. Ferrous and ferric ions in the VNIR region produce good absorption properties.

Carbon shales have no special adsorption properties. It is also pure siliceous sandstone without special adsorption properties. Sandstone usually has iron oxide veins that produce spectral properties at 870 nm. Limestones and rocks with a calcareous composition are characterized by carbonate adsorption bands (1990 and 2350 nm) and are more pronounced in the range of 2350 nm (Rangzan et al. 2018).

Metamorphic Rocks

Extensive adsorption due to ferrous iron ions is a predominant feature in rocks such as tremolite. Hydroxyl and water bonds (1400 and 1900 nm) are present in schists and marbles and quartz. Carbonate bands (1990 and 2350 nm) are characteristic of marble.

Alteration Zones

These zones have important indications for the exploration of montmorillonite, kaolinite, sericite, muscovite, biotite, chlorite, epidote, pyrophyllite, alunite, zeolite, quartz, albite, goethite, hematite, broomstone, carbonates, serpentine, actinolite (Clark and Roush 1984).

These alteration minerals can be divided into 5 groups:

(A) Quartz + feldspar (quadrilateral silicates) which show no spectral properties in the amplitude—VNIR SWIR and show only an increasing trend in reflection (reflection)
(B) Clay (sheet silicates), characterized by absorption bands at 2100–2300 nm have become.
(C) Carbonates have spectral properties at 2350 nm.
(D) Hydroxyl and water produce adsorption properties at 1400 and 1900 nm.
(E) Iron oxides, their spectral properties can occur in the VNIR region. Relative rate this property can be changed and these changes can be caused by changes in the spectral reaction.

7.2.5 Practical Examples of Detrital Sediment Spectroscopy Studies

Since the main purpose of this book is to study sediments and sedimentary rocks, in the following section, an example of spectroscopic studies performed on detrital sediments of Aghajari Formation in southwestern Iran is presented. The aim of this study is to investigate the kaolinite and montmorionite of Aghajari sandstones for possible uranium exploration with the help of FieldSpec 3 spectrometer (Zargari et al. 2014).

Identifying and defining the spectral properties of minerals and rocks using modern methods of remote sensing in the field of conventional methods is very important.

Among the methods that will reduce the need for field studies is spectrometry. Non-Imaging spectrometry can be used in lab environment such as Fieldspec3.

Presence of clay minerals in Aghajari sandstone of the northern part of Khuzestan province, with dominant reducing conditions due to the presence of hydrocarbon reservoirs, can be of promising nature.

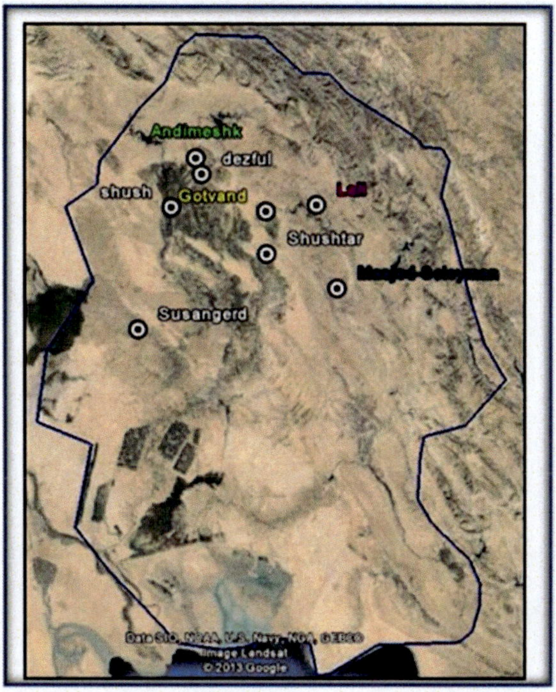

Fig. 7.11 Location of the study area in the project

Many clay minerals adsorb metallic minerals such as uranium.

The clay minerals present in the Aghajari sandstone were studied by calculating the depth of the absorption spectra with Fieldspec3 at wavelengths around 0011, 0011 and 0011 nm and compared with that of pure kaolinite and montmorillonite.

The abundance of 1.0–1.0% grade uranium in sandstone formation may indicate meaningful relation to the absorption depth of the minerals Kaolinite–Montmorillonite.

7.2.5.1 Study Area

The region in question is located in the north of Khuzestan province in southwestern Iran (Fig. 7.11).

7.2.5.2 Research Methods

The studies conducted in this research are divided into two parts: geological studies and book studies (analysis and preparation of samples).

Geological and Petrological Studies of the Area

The study area is located in Khuzestan plain (southwest of Iran) and includes sandstone units of Aghajari Formation (from the detrital formation located in the Zagros Basin of Iran). Lithologically, these sandstones alternate with marl and silt.

Data Analysis and Preparation

At this stage, the samples collected from the study areas and the samples of pure kaolinite and montmorillonite, after being transferred to the spectroscopy laboratory of Shahid Chamran University of Ahvaz, Iran, were examined by the field spectroscope 3. In these studies, ENVI4.8 and RPISM software have been used.

Removal of Continuum and Absorption Band Depth

In the spectra obtained from different materials in this research, the absorption of the electromagnetic spectrum is investigated by two factors, which include continuum and absorption shape.

A continuum is the upper range of the reflective curve and is a subset of the larger absorption shape of matter. The absorption form is a factor that is investigated by the absorption depth parameter.

The following equation is used to calculate the absorption depth (Roger et al. 2003):

$$D = Rc - Rb/Rc = 1 - (Rb/Rc)$$

In this equation D (absorption depth), Rb (maximum absorption reflection), Rc (continuous reflection).

Note that Rb and Rc are calculated at the same wavelength.

Absorption Band of Kaolinite and Montmorillonite

Two-band kaolinite is the predominant adsorption band around 1400 nm due to the presence of hydroxyl ions and 2200 nm due to the combination of OH with magnesium and aluminum in the spectral curve. Montmorillonite also shows an attractive shape at about 1400, 1900 and 2200 nm due to its similar composition to kaolinite (Clark 2003).

Uranium Ore Sand Deposits

These deposits are one of the most important sources of uranium in the world. These deposits are found in continental and generally river sediments, which are mostly left in dry conditions. These deposits contain 1000–10,000 tons of uranium with a grade of 0.1–0.2%. Uranium is often adsorbed to the clays superficially with a decrease in EH (Shahabpour 2007).

7.2.5.3 Results of This Research

Depth calculation and relationship with purity (carat).

The absorption depth was calculated at about 1400, 1900 and 2200 nm with PRISM software. Then, by applying the ratio between the adsorption depths of sandstone samples and pure kaolinite and montmorillonite samples (as a control) according to the following equation, first the relationship between the adsorption depth and the purity percentage of clay minerals is obtained.

Percentage of clay mineral sample = (depth of sandstone adsorption * 100)/depth of mineral clay adsorption depth (Figs. 7.12 and 7.13).

After calculating the purity percentage of clay minerals in the Aghajari sandstone diagram, the possible amount of uranium was obtained using the following equation:

Sandstone grade = (sandstone ore grade * percentage of clay mineral purity in the sample)/100

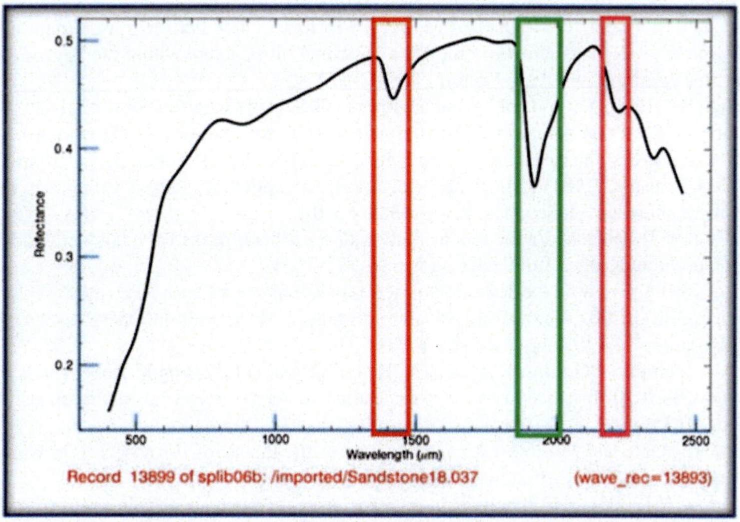

Fig. 7.12 The absorption depth was calculated at about 1400, 1900 and 2200 nm with PRISM software

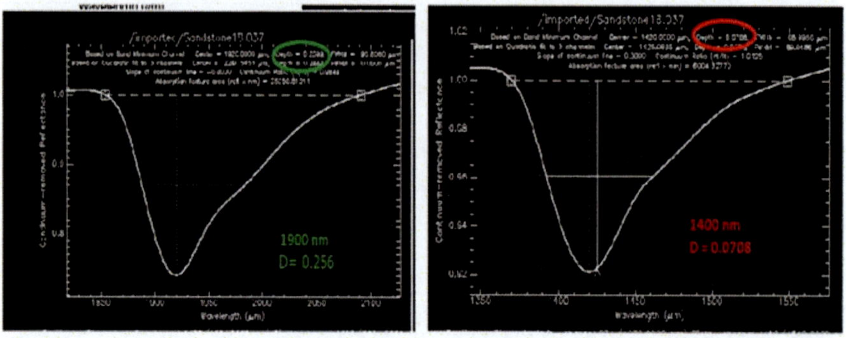

Fig. 7.13 Analysis of one of the sandstone samples of Aghajari Formation in southwestern Iran and calculation of absorption depth at 1400, 1900 and 2200 nm wavelengths (by PRISM software)

After calculating the potential grade of uranium, the maximum grade of 0.135% for sandstones containing kaolinite and 0.115%, which is close to the global grade of uranium ores (0.2–0.1%).

References

Askandari Nejad F, Morteza FA, Musa FA (2015) Investigation of flood potential of Nimrud basin and its effect on socio-economic characteristics of the region and provide solutions for its management. J Quant Geomorphol Res, Fourth Year 1:75–89

Alizadeh A, Hormozi H, Moghadam M, Seraj M (2020) DEM-derived geomorphic indices for assessment of tectonic activity at the Dara anticlinal oil structure within the Zagros fold-thrust belt, southwestern Iran. Arab J Geosci 13:192–212

Burrough PA (1990) Methods of spatial analysis in GIS. Int J Geogr Inf Syst 4:221–223

Casalbore D, Passeri F, Tommasi P (2020) Small-scale slope instability on the submarine flanks of insular volcanoes: the case-study of the Sciara del Fuoco slope (Stromboli). Int J Earth Sci

Clark RN, Roush L (1984) Reflectance spectroscopy quantitative analysis techniques for remote sensing applications. J Geophys Res 89:6329–6340

Costa JE (1983) Palaeohydraulic reconstruction of flash-flood peaks from boulder deposits in the Colorado Front Range. Bull Geol Soc Am 94:986–1004

Feng Z (2019) A review on the definitions of terms of sedimentary facies. J Paleogeography 8:32–48

Forman E, Saul G (2001) The analytical hierarchy process—an exposition. Oper Res 49(4):469–487. https://doi.org/10.1287/opre.49.4.469.11231

Forzieri G, Miralles DG, Ciais P, Alkama R, Ryu Y, Duveiller G, Zhang K, Robertson E, Kautz M, Martens B (2020) Increased control of vegetation on global terrestrial energy fluxes. Nat Clim Chang 10:356–362

Ghazanfari M, Rezaei M (2006) An introduction to the theory of fuzzy sets. Iran University of Science and Technology Publications, p 762

Ghodsipour H (2006) AHP hierarchical analysis process. Amir Kabir University of Technology Publications, p 638

Gomeh Z, Rangzan K, Nazari Saman A, Ghodosi J (2014) Evaluation of green space changes with respect to landscape index and remote sensing in capital city of Karaj, vol 67, no 3. Serial Number 3, pp 323–331

Gupta S (2002) Investigations of micro-stress and phase transition in sol–gel-derived multideposited coatings of barium titanate using Raman spectroscopy. J Raman Spectrosc 33(1):42–49

Hagos M, Gebreyohannes T, Amare K (2020) Tectonic link between the Neoproterozoic dextral shear fabrics and Cenozoic extension structures of the Mekelle basin, Northern Ethiopia. J Earth Sci

Hunt GR, Salisbury JW, Lehnoff CJ (1971) Visible and near infrared spectra of minerals and rocks: 111 oxides and oxyhydroxides. Modern Geol 2:195–205

Jensen JR (2015) Introductory digital image processing (a remote sensing perspective), Pearson Education, 659 pp

Liu H, Meng J, Banerjee S (2017) Estimation of palaeo-slope and sediment volume of a lacustrine rift basin: a semi-quantitative study on the southern steep slope of the Shijiutuo Uplift, Bohai Offshore Basin, China. J Asian Earth Sci 147:148–163

Malkian A, The fallen of Khuzani A, Ashournejad Gh (2012) Zoning potential flooding of Akhtarabad watershed using fuzzy hierarchical analysis method. Q J Nat Geogr Res 82:131–152

Nagender-Nath B (2020) Dynamics of the earth system evolution-processes and interactions. Soc Earth Sci 127–136

Omidvar K, Kianfar A, Askari Sh (2010) Flood potential zoning of Kanjancham Basin. Q J Nat Geogr Res 72:73–90

Piao S, Wang X, Park T, Chen C, Lian X, He Y, Bjerke JW, Chen A, Ciais P, Tømmervik H (2020) Characteristics, drivers and feedbacks of global greening. Nat Rev Earth Environ 1:14–27

Pourmorad S, Harami R, Solgi A, Ale Ali M (2021a) Sedimentological, geochemical and hydrogeochemical studies of alluvial fans for mineral environment purposes, lithology and mineral resources, vol 56, pp 89–112

Pourmorad S, Mostofa KhMG, Li SL, Liu CQ, Moein Z (2021b) Drilling engineering (geology and mud engineering). Lambert publication, Germany, p 470

Qanawati A, Karam A, Aghaalmkgni M (2011) Efficiency of hierarchical analysis method in flood studies. Geogr Q 31:255–275

Rangzan K, Mirzaei R, Ashrafzadeh M (2018) Potential distribution modelling of wildlife species based on ecological knowledge of local communities compared with machine learning methods: a case study of Gazella subgutturosa in Mishdagh protected area. J Nat Environ 70(4):893–906

Salavati A, Haghshenas H, Ghadirifaraz B, Laghaei J, Eftekhari G (2016) Applying AHP and clustering approaches for public transportation decision making: a case study of Isfahan City. J Public Transp 19(4):3

Sanchez-Nuñez JM, Macías JL, Saucedo R (2017) Geomorphology, internal structure and evolution of alluvial fans at Motozintla, Chiapas, Mexico. Geomorphology 230:1–12

Sime CH, Demissie TA (2020) Surface runoff modeling in Ketar watershed. Ethiop J Sediment Environ 5:151–162

Shoorangiz M, Sarkarinejad K, Nourbakhsh A (2020) Tectonic implication of quantitative microfabric analyses of quartz c-axis development within the Tutak gneiss dome, Zagros hinterland fold-and-thrust belt. Int J Earth Sci (GeolRundsch) 109:127–144

Zargari Z, Rangzan K, Saberi A (2014) Investigation of Kaolinite-Montmorillonite absorption depth of Aghajari formation sandstone in Khuzestan for possible exploration of Uratium by Fieldspack 3, In: 32nd international specialized conference on earth sciences, Iran

Zhang Y, Ye W, Ma C (2020) Middle to Late Holocene changes in climate, hydrology, vegetation and culture on the Hangjiahu Plain, southeast China. J Paleolimnol

Chapter 8
Basin Analysis and Modeling

Abstract Sedimentary basin analysis is one of the main and final factors of any geochemical and sedimentological studies (Pourmorad and Jahan 2021). In these studies, all the obtained results should be carefully analyzed, and a correct interpretation of all these results and finally the final sedimentary model should be introduced. In this section, as an example, the results of the studies of alluvial fans of southwestern Iran are summarized, and finally a final model is introduced.

Keywords Modeling · GB · SG · SB · Dezful model

8.1 Introduction

Basin modeling is the term broadly applied to a group of geological disciplines that can be used to analyze the formation and evolution of sedimentary basins, often but not exclusively to aid the evaluation of potential hydrocarbon reserves (Lee et al. 2019; Hantschel and Kauerauf 2009; Duppenbecker and Eliffe 1998; Lotveit et al. 2019).

Basin modeling enables geoscientist to investigate the dynamics of sedimentary basins and their associated fluids to determine if the past conditions were appropriate to fill potential reservoirs with hydrocarbon and preserve the potential reservoirs (Allen 2013; Al-Hajari et al. 2009; Grunnaleite and Mosbron 2019).

Basin modeling applies mathematical algorithms to seismic, stratigraphic, paleontological, petrophysical, well log and geologic data to reconstruct the evolution of sedimentary basins (Hantschel and Kauerauf 2009; Bjorlykke 2010).

Model building includes constructing a structural model and identifying the chronology of deposition and physical properties of each layer. The necessary inputs for model buildings can be explained under these substages: geometry and stratigraphy, timing and deposition, geochemical data and boundary conditions (Pourmorad et al. 2021a, b, c; Cathles and Fjeldskaar 2020).

Since the preparation of subsurface sediment models has been widely discussed in articles and books and less has been discussed about the preparation of surface sediment models, this chapter has tried to provide a suitable model for the preparation of this type of study.

S. Pourmorad and A. Mohanty, *Alluvial Fans in Southern Iran*, Advances in Geographical and Environmental Sciences, https://doi.org/10.1007/978-981-19-2045-5_8

In this chapter, first, introductions and preliminary studies for the preparation of the sedimentary model have been introduced and then the sedimentary model of the study area has been introduced.

8.2 Basin Assessment

In order to provide a comprehensive sediment model that can be a correct and ideal model for the study area, accurate assessment of the area in terms of all sedimentary, stratigraphic, geochemical and tectonic properties is very important (Pourmorad and Jahan 2021;Cathles 2019).

In this section, we have tried to provide all the necessary information to provide a sedimentary model of the study area to be used as a model in this type of study.

The study of alluvial fans in southwestern Iran shows that the most abundant particle size in the upstream part of Dezful alluvial fans is −4 to −2 Phi (small to large gravels) (Pourmorad et al. 2021a, b, c). In Ramhormoz region −3 to −4 Phi (medium to large gravels), in the middle part in Dezful region 1 to −1 Phi (very coarse sands) and in Ramhormoz region in the size of 0 to 1 Phi (sand in the downstream part in both areas is larger than 4 Phi (silt and clay)) (Pourmorad et al. 2021a, b, c).

Statistical studies conducted in these areas showed that the average and mean in the samples studied in the opposite parts of these alluvial fans are different due to the distance from the top and with the distance from the origin to the downstream gradually moderate and the average tends to be finer grain particles. In addition, the sorting rate in Dezful region is good to moderate in the upper part, weak in the middle part and moderate in the downstream part, but in Ramhormoz region, the sorting increases from the top to the downstream so that at the peak sorting is weak and downstream sorting is good to moderate.

Examination of the amount of skewness and elongation of sediments in the study areas in southwestern Iran showed that the amount of skewness in both study areas at the top toward fine grains and in the middle part is almost symmetrical and in the downstream part toward coarse grains. The amount of elongation of the specimens in the upper part of Dezful is moderate to elongate in Ramhormoz region, in the middle part of Dezful it is elongated and in Ramhormoz region it is moderate to elongated, and in the lower part of both regions it is medium to wide.

According to studies, rock facies in the study areas are divided into three categories, gravel (including Gmm, Gmg, Gcm, Gp and Gh facies), sandy (including SmSp, Sh facies) and mud (including FM, Fi facies). Also, five structural elements CH, GB, SG, SB and FF have been identified in the mentioned areas.

Estimation of current strength using part size in the studied areas shows that the highest flow velocity in the two study areas belongs to Ramhormoz area with maximum current power (591.82 w) and maximum discharge rate) 166.32 m^3s^{-1}).

According to these data, Gmm and Gcm facies in both study areas are associated with high values of power and flow velocity, while Gh and Gp facies in both regions have the lowest values of calculated velocity and power. In addition, according to studies, the maximum annual discharge, which corresponds to the maximum thickness of oblique groups, belongs to Ramhormoz area (203.21 m^3s^{-1}).

Petrographic studies in these areas show that the source of alluvial fan sediments in the north of Khuzestan province (southwest of Iran) are Aghajari and Bakhtiari formations and in the central part of Khuzestan province (southwest of Iran) they are Gachsaran and Bakhtiari formations.

The results of geochemical analysis of the studied samples show that the most abundant oxides in both Dezful and Ramhormoz are calcium oxide (CaO) and magnesium oxide. Also, the most abundant rare earth elements in Dezful region include Ce, La, Nd and Y, respectively, but in Ramhormoz region, they include Ce, La, Y and Nd, respectively. In addition, in Dezful and Ramhormoz regions, the abundance of heavy metals includes Zn, Pb, Cu and Cd, respectively.

Preliminary weathering studies of sediments using CIA and CIW indices show the long distance of sediment transport from the source area and the humid conditions and climate of the source area (Zagros highlands) for these samples.

Study of the tectonic position of sedimentary samples based on geochemical results shows that sedimentary samples of Dezful region are mainly in the Active continental margin (probably due to Sanandaj-Sirjan zone) and to a lesser extent in the Continental island Arc (probably due to it being located in Dokhtar Urmia zone), but the samples of Ramhormoz region are mainly located in Active continental margin region.

According to the diagrams of Hayashi et al. (Hayashi et al. 1997), the sediments of Dezful region are generally of felsic origin, but in Ramhormoz region, these sediments are probably of moderate felsic origin.

Pollution study based on Igeo index shows that the average pollution caused by Cu, Zn, Pb, Ni and V elements in Dezful region is relatively low, but in Ramhormoz region vanadium (v) and to some extent Ni-induced pollution. And Zn is high, which is more likely due to oil drilling fluids in upstream areas such as the Mamatin oil field.

The data obtained from these mineralogical studies by the XRD method show that chlorite and illite are the most abundant clay minerals in the studied deposits, especially in Dezful region. In Dezful region, calcite, quartz, feldspar and dolomite are among the non-clay minerals in these deposits, and in Ramhormoz region, in addition to these minerals, gypsum is seen as another non-clay mineral.

Groundwater hydrochemical studies of the studied areas show that the pH level in both Ramhormoz and Dezful areas is normal and is in the range of 6.5–7.9. Also, the amount of electrical conductivity in Ramhormoz region is more than Dezful region. This electrical conductivity is more in the water samples of Ramhormoz region due to the presence of Gachsaran Formation and consequently the presence of salt, which is the main factor in increasing electrical conductivity. In addition, according to these studies, the amount of sulfate, sodium and calcium in Ramhormoz is much higher than in Dezful.

Tectonic studies conducted in the study areas show that the most important tectonic factors in Ramhormoz region include Ramhormoz thrust fault, Persian anticline and in Dezful region including Dezful fault, mountain front fault. Studies conducted in the study areas in terms of seismicity show that Andimeshk and Ramhormoz study areas are in relatively low seismic hazard ranges.

From the total information obtained, it can be concluded that both Ramhormoz and Dezful regions have almost the same flood risk because in Dezful region, despite the greater impact of factors such as rainfall, rainfall intensity, slope and density of faults. Due to lithology with good permeability, we see a decrease in the volume of surface runoff.

8.3 Sedimentary Model

Investigation of grain size characteristics, sedimentary facies and grain size changes in the study areas indicates that the sediments of the alluvial fans have almost the same sedimentation conditions.

In Ramhormoz region, the mentioned sediments are divided into three proximal, medial and distal parts based on changes in grain size and changes in the slope of the region by transporting them from the heights of Gachsaran Formation to the downstream parts. In the mentioned area, it is located at the top of Gachsaran Formation, which is located on Aghajari Formation with Ramhormoz thrust, and in the eastern and southeastern part of the region, Lahbari section is driven on Aghajari Formation. In these areas, changes in river flow intensity have caused many changes in the characteristics of sedimentary structures in this area, especially in the proximal part. In the medial part, due to the distance from the apex of the alluvial fan, the intensity of the particle currents is reduced and as a result, we see less facies diversity than the proximal part. In the distal part in this region, due to its confinement with the Lahberian thrust, we see its expansion in the southwestern parts and its limitation in the southern and southeastern parts (Fig. 8.1).

The alluvial fans in Dezful region flow from the heights of Bakhtiari Formation to the downstream parts. In these areas, the role of particle currents in most of the alluvial fans is much more significant than the role of river currents. In this region, sedimentation from the apex to the distal part is associated with gradual changes in facies (Fig. 8.2).

Finally, based on field evidence and study of lateral and vertical changes of sedimentary facies, siliceous-detrital deposits of the mentioned areas under the influence of river flows and small streams from the heights of the areas to the lower parts. The hands have flowed.

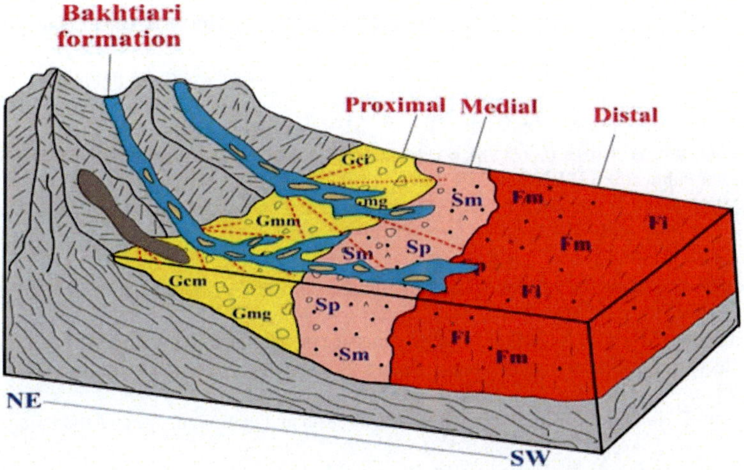

Fig. 8.1 Sedimentary model of alluvial fans of Ramhormoz study area, southwest of Iran

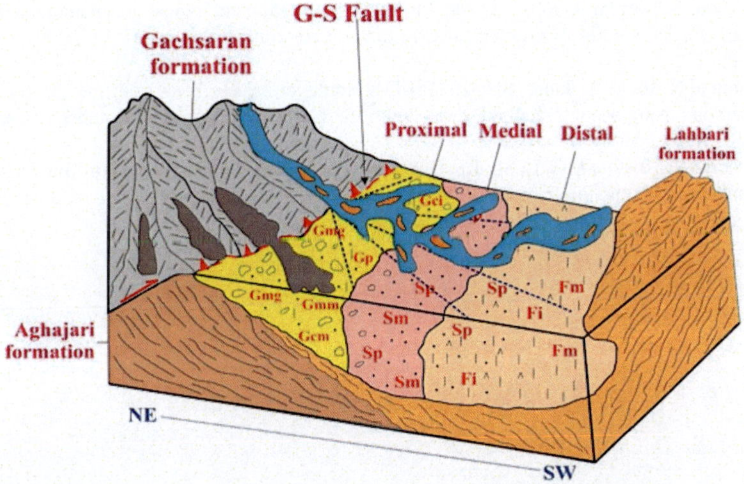

Fig. 8.2 Sedimentary model of alluvial fans of Dezful study area, southwest of Iran

References

Al-Hajari M, Derks J, Fuchs T, Hantschel T, Kauerauf A, Neumaier M, Schenk O, Swientek O, Tesse NN, Welte D, Wygrala B, Kornpihi D, Peters K (2009) Basin and petroleum system modeling. Oilfield Rev 21(2). Copyright Schlumberger

Allen PA, Allen JR (2013) Basin analysis: principles and application to petroleum play assessment, 3rd edn. Oxford, United Kingdom, p 760

Bjorlykke K (2010) Petroleum geoscience. Springer, Berlin, Heidelberg, pp 349–360. https://doi.org/10.1007/978-3-642-02332-3_15

Cathles L (2019) On the processes that produce hydrocarbon and mineral resources in sedimentary basins. Geosciences 9:520

Cathles L, Fjeldskaar W (2020) A summary of "future advances in basin modeling: suggestions from current observations, analyses and simulations". Geosciences 10(12):506. https://doi.org/10.3390/geosciences10120506

Duppenbecker SJ, Eliffe JE (1998) Basin modelling: practice and progress. Geological Society Special Publication. ISBN 1-86239-008-8

Lee EY, Novotny J, Wagreich M (2019) Subsidence analysis and visualization—for sedimentary basin analysis and modelling. Springer. ISBN 978-3-319-76423-8. https://www.springer.com/gp/book/9783319764238

Lotveit IF, Fjeldskaar W, Sydnes M (2019) Tilting and flexural stresses in basins due to glaciations an example from the Barents sea. Geosciences 9:474

Grunnaleite I, Mosbron A (2019) On the significance of salt modelling example from modelling of salt tectonics, temperature and maturity around salt structures in southern north sea. Geosciences 9:363

Hantschel T, Kauerauf A (2009) Fundamentals of basin and petroleum systems modeling, pp 103–105. https://doi.org/10.1007/978-3-540-72318-9

Hayashi KI, Fujisawa H, Holland HD, Ohmoto H (1997) Geochemistry of approximately 1.9 sedimentary rocks from northeastern Labrador, Canada. Geochim Cosmochim Acta 61:415–437

Pourmorad S, Jahan SH (2021) A model for comprehensive studies of alluvial fan deposits, case study: Ramhormoz Mega-fan in southwest Iran. J Earth Sci Clim Chang 12(Issue 3)

Pourmorad S, Harami R, Solgi A, Ale Ali M (2021a) Sedimentological geochemical and hydro-geochemichal studies of alluvial fans for mineral environment purposes. Lithol Miner Resour 56:89–112

Pourmorad S, Jokar A, Jahan SH (2021b) Determination of key beds from the cap rocks of oil reservoirs using a novel method, case study: the Gachsaran Formation, southwest Iran. Lithol Miner Resour 56(6):559–578

Pourmorad S, Mostofa MG, Li SL, Liu CQ, Moein Z (2021c) Oil and gas well drilling engineering. Lambert Academic Publishing, Germany, 480 p

Appendix

Attached image

S. Pourmorad and A. Mohanty, *Alluvial Fans in Southern Iran*, Advances in Geographical and Environmental Sciences, https://doi.org/10.1007/978-981-19-2045-5

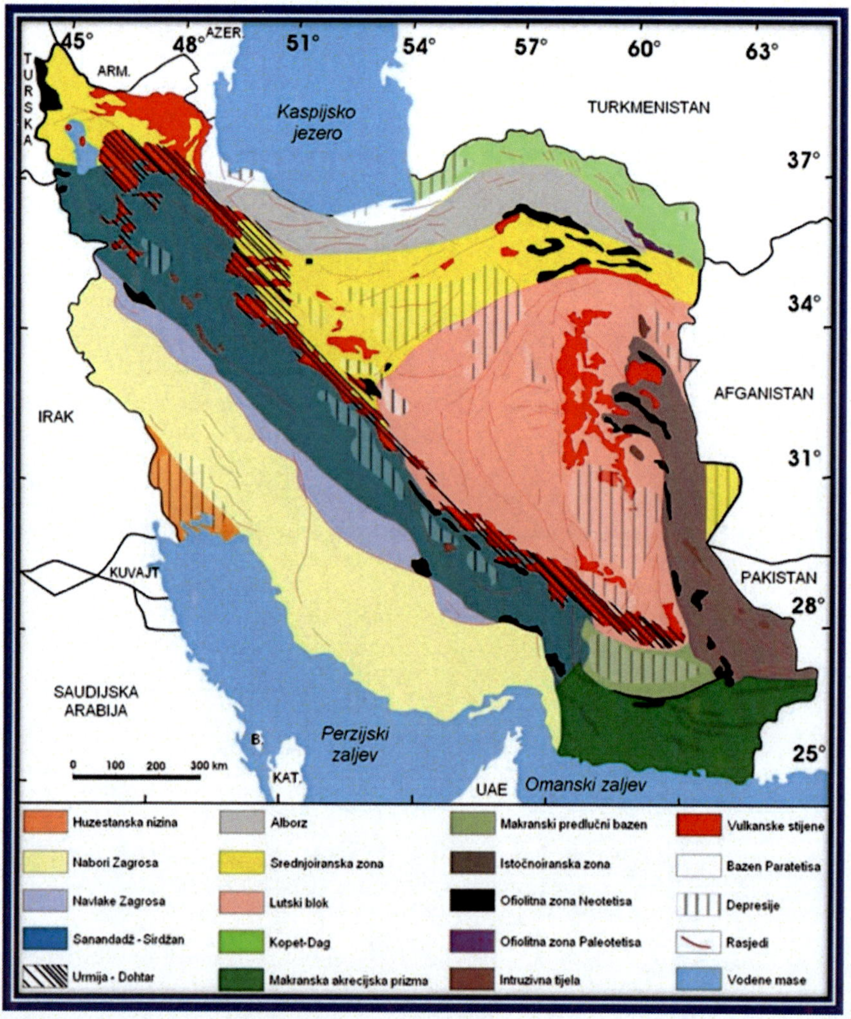

Fig. A.1 Geological_zones_of_Iran_(Alavi, 2012)

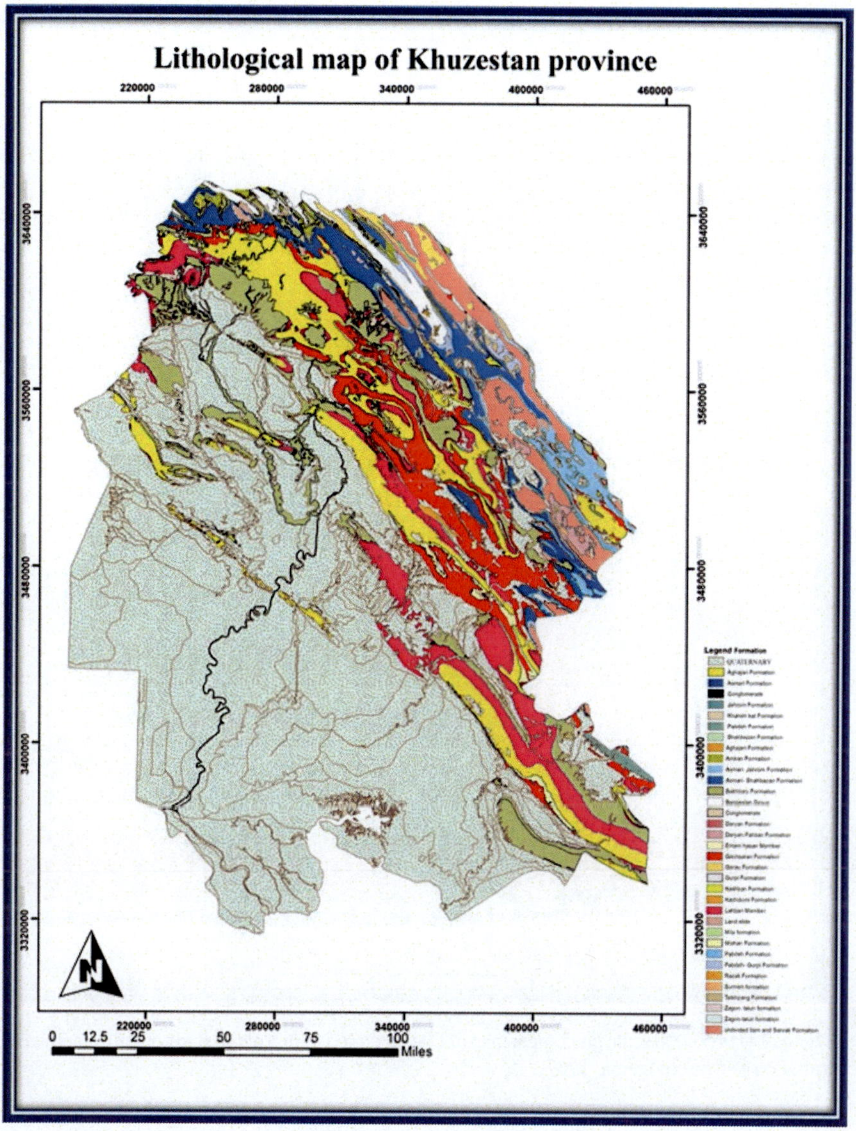

Fig. A.2 Geological map of Khuzestan province in southwestern Iran, the site of case studies (Geological Survey of Iran (Ahvaz center), 2019

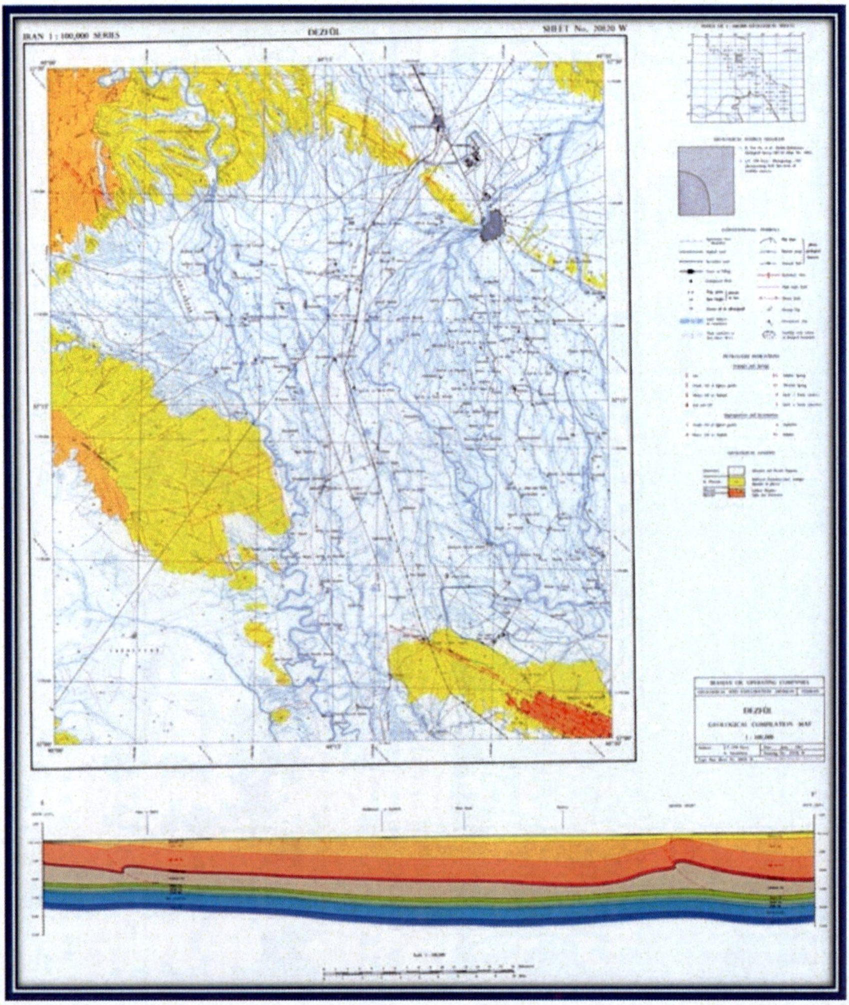

Fig. A.3 1: 100,000 scale geological map of Dezful region as one of the study areas in southwestern Iran, Perry and Setudehnia, 1967

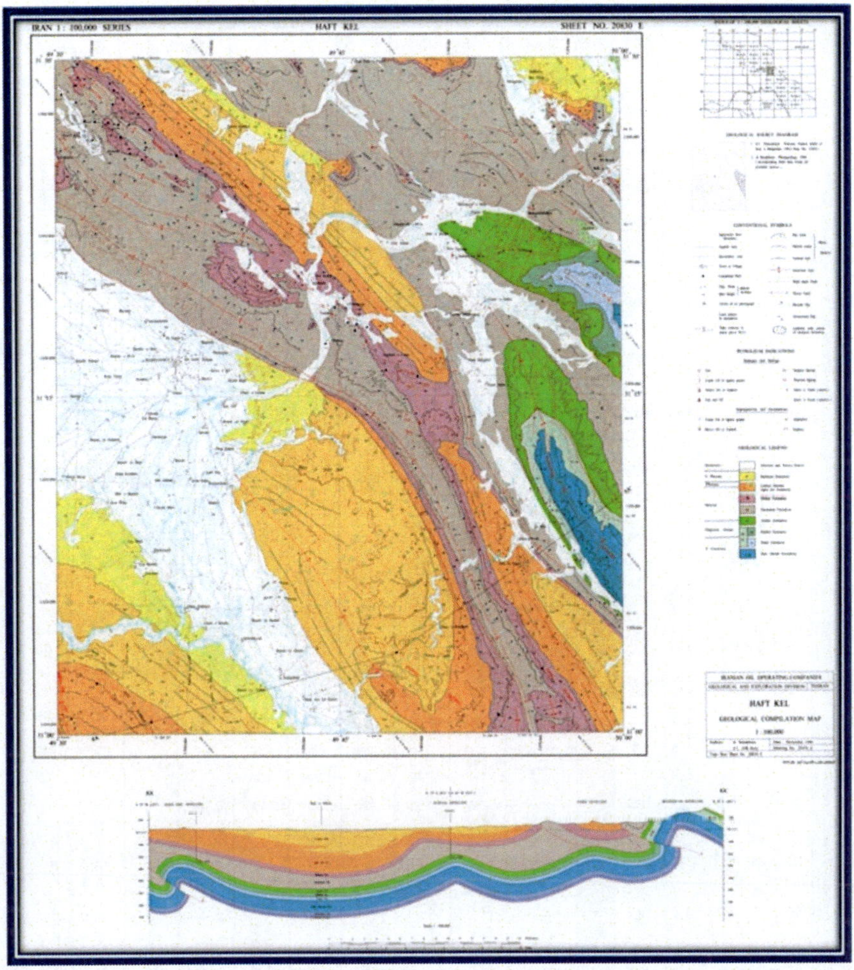

Fig. A.4 Geological map of Haftgel with a scale of 1: 100,000 as one of the study areas in southwestern Iran, (Ramhormoz study area in the center of the map), Setudehnia and Perry, 1966

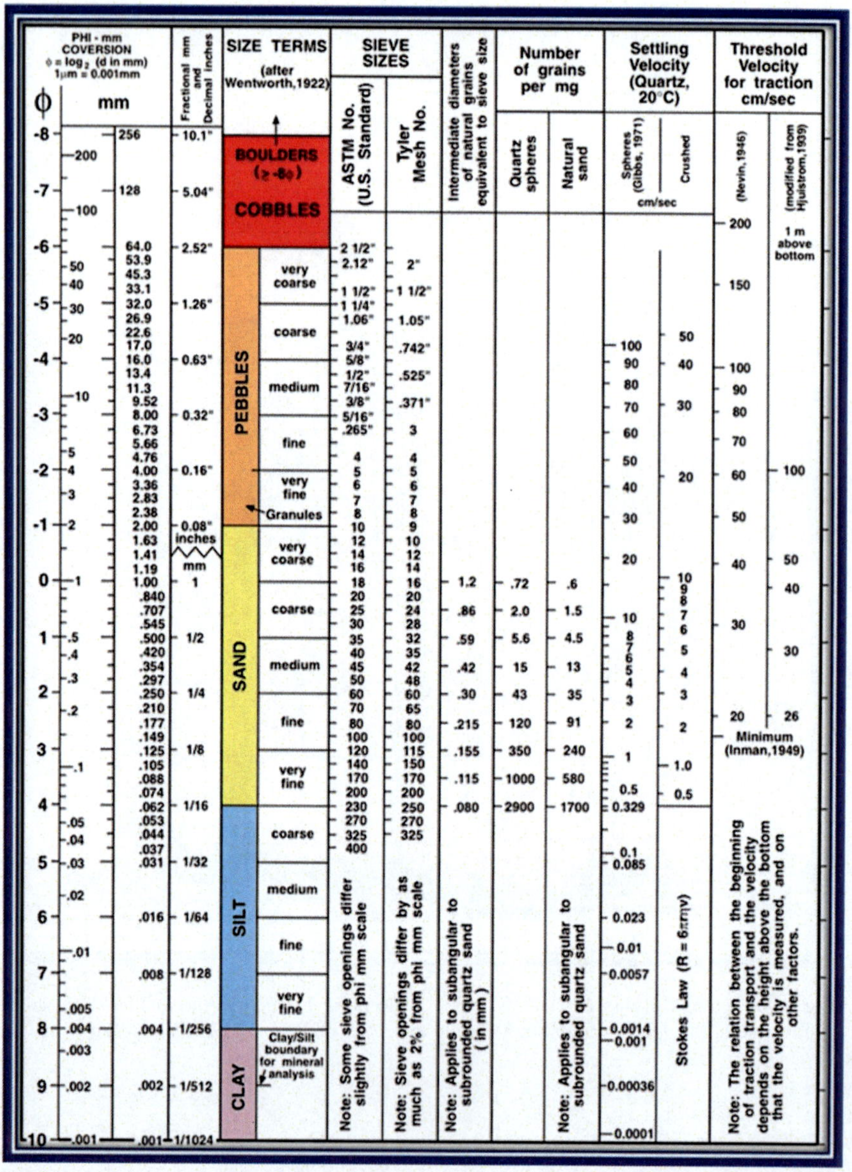

Fig. A.5 Wentworth grain size chart from United States Geological Survey Open-File Report 2006–1195

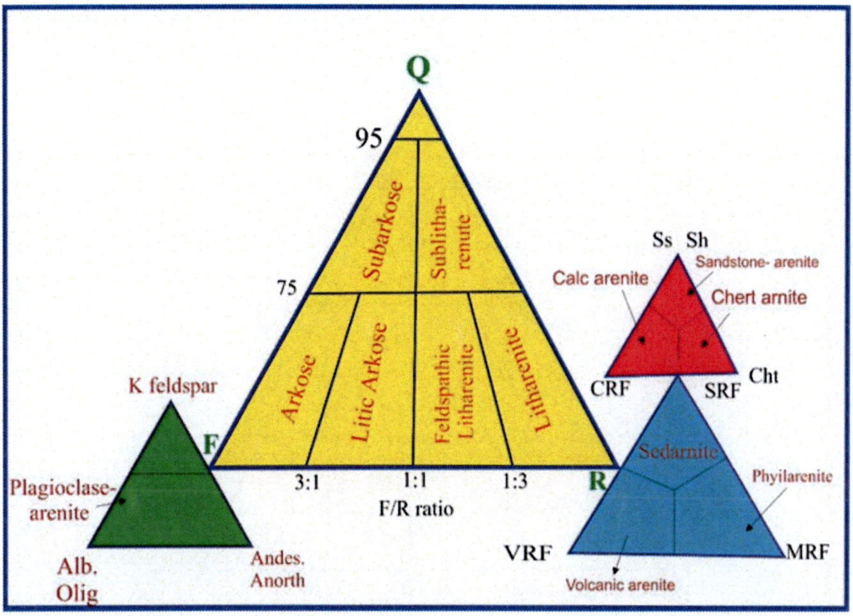

Fig. A.6 Classification of folk sandstones, 1974

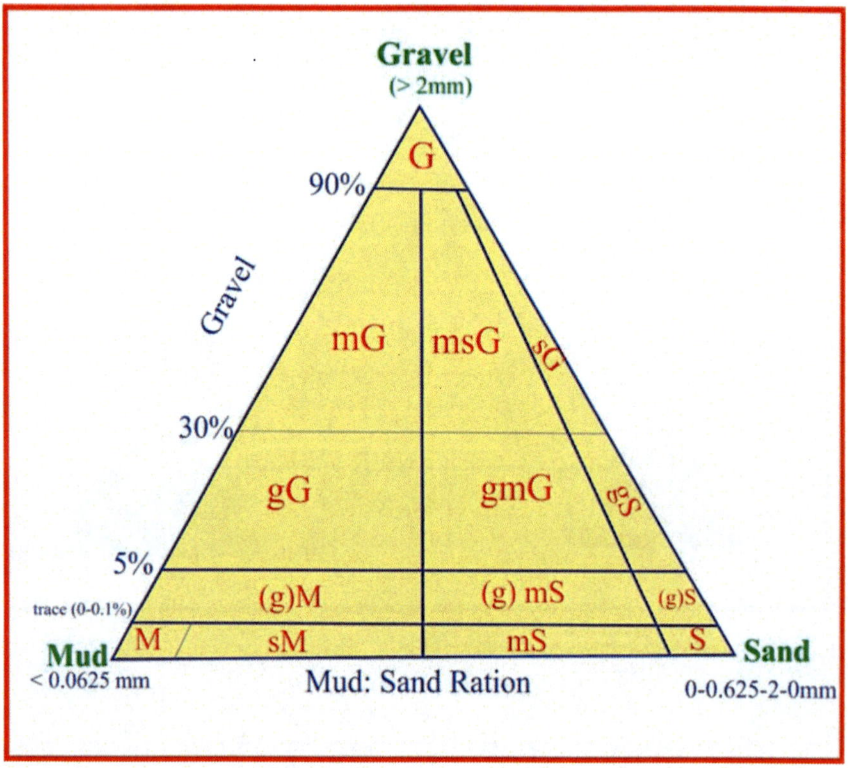

Fig. A.7 Textural classification of gravel-bearing sediments and sedimentary rocks (Folk, 1970)

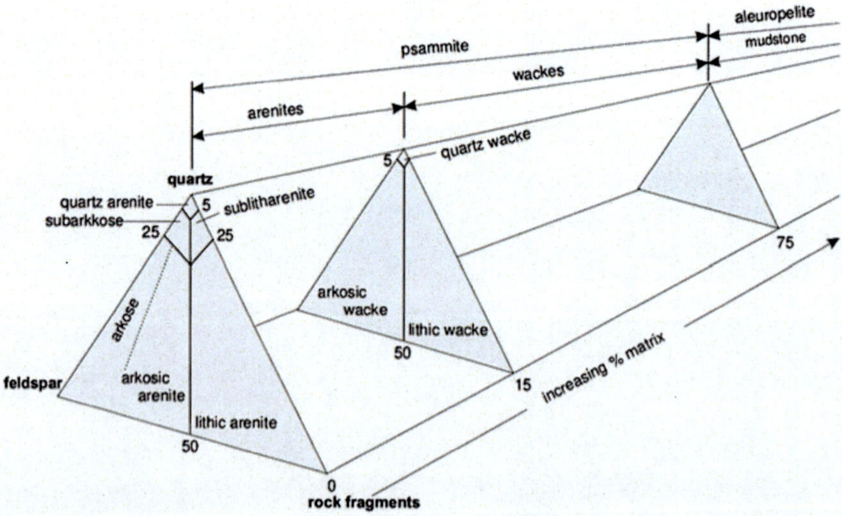

Fig. A.8 Classification of the Sandstone (modified from Pettijohn_1975)

Facies Code	Facies Description	Interpretation
Gmi	Massive intraformational conglomerates	Lag deposits, longitudinal bars
Gp	Planar cross-bedded conglomerates	Transverse bedforms
Gh	Clast-supported horizontally stratified cross conglomerates	Longitudinal bedforms, lag deposits, sieve deposits
SGp	Planar cross-bedded gravelly sandstone	Transverse and linguoid gravelly sand bedforms, (2-D) dunes and bars.
Sp	Planar cross-bedded sandstone	Transverse and linguoid bedforms (2-D) dunes
St	Trough cross-bedded sandstone	Sinous crested and linguoid (3-D) dunes
Sh	Horizontally bedded sandstone	Plane bed flow (critical flow)
Sm	Massive sandstone	Sediment gravity flow deposits
Sr	Ripple cross-Laminated sandstone	Ripples (lower flow regime)
Fr	Ripple cross-Laminated mudstone	Ripples (lower flow regime)
Fl	Laminated mudstone	Overbank, abandoned channel. Or waning flood deposits
Fm	Massive mudstone	Overbank, abandoned channel. Or drape deposits

Fig. A.9 Lithofacies types and codes modified from Miall (2010)